Understanding Building Failures

Third edition

James Douglas and Bill Ransom

Taylor & Francis
Taylor & Francis Group

LONDON AND NEW YORK

First edition 1981 by E & FN Spon
Second edition 1987
Reprinted 1990, 1993, 1995, 1996
Reprinted 2001
by Spon Press

This edition published 2007
by Taylor & Francis
2 Park Square, Milton Park, Abingdon, Oxon OX14 4RN

Simultaneously published in the USA and Canada
by Taylor & Francis
270 Madison Ave, New York, NY 10016

*Taylor & Francis is an imprint of the Taylor & Francis Group, an informa
business*

© 1981, 1987 Bill Ransom; 2007 James Douglas and W.H. (Bill) Ransom

Typeset in Sabon by
Integra Software Services Pvt. Ltd, Pondicherry, India
Printed and bound in Great Britain by
The Cromwell Press, Trowbridge, Wiltshire

British Library Cataloguing in Publication Data
A catalogue record for this book is available from the British Library

Library of Congress Cataloging in Publication Data
Douglas, James.
    Understanding building failures / James Douglas and W.H. Ransom. --
    3rd ed.
        p.   cm.
    Includes bibliographical references and index.
    ISBN 0-415-37082-5 (hc : alk. paper) -- ISBN 0-415-37083-3 (pbk. : alk.
    paper) 1. Building failures. 2. Buildings--Defects. 3. Buildings--Repair
    and reconstruction. I. Title.
    TH441.D68 2006
    690'.21--dc22
    2006010403

ISBN10: 0–415–37082–5 (hbk)
ISBN10: 0–415–37083–3 (pbk)

ISBN13: 978–0–415–37082–0 (hbk)
ISBN13: 978–0–415–37083–7 (pbk)

Contents

Figures

Tables

Preface to third edition

Despite the growing awareness as to many of their common causes and consequences, failures still seem to bedevil the building industry in the United Kingdom and elsewhere in the developed world. This revised edition addresses the reasons for this problem and offers some guidance on how it can be tackled. Its title has been modified to emphasise the need to understand the circumstances behind the occurrence and investigation of defects in buildings. This understanding will address the following 'journalist questions':

- How? (the agencies and mechanisms)
- Why? (the aetiology and diagnosis)
- When? (the timing and longevity)
- Where? (the location and position)
- Who? (culpable and responsible person/s).

Why is the need to understand defects of such importance for construction professionals such as building surveyors, architects and others? There are three main reasons for the increase in demand for this knowledge. First, there is the ambitious target of zero defects as articulated in the British building industry's key performance indicators (KPIs). This will only be attained, if at all, if professionals, contractors and operatives are made aware of the causes and consequences of building failures.

Second, the need to reduce repair and legal costs resulting from defects remains high. This is a disruptive as well as expensive process for clients and others involved.

Third, the recent introduction of the Home Condition Report or single house survey has placed additional demands on building practitioners. Building surveyors and other construction professionals undertaking this service need to have a good knowledge of building construction as well as an ability to identify and diagnose relevant defects.

Much of the insights related to diagnostics in this edition, such as decision making and critical thinking, are derived from the medical and nursing professions. The primary reason for this is that the diagnostics experience and research base of these health-care disciplines is well in advance of that for construction.

The growth in the number of textbooks dealing with defects has been slow but steady. For many years the book by Eldridge (1974) was the standard reference text on building failures in the United Kingdom. That all changed in the 1980s with the increasing number of books on building surveys and inspections (e.g. Mika 1989, Melville and Gordon 1993, and Bowyer 1994). The works by the NBA (1983), Cook and Hinks (1992), Son and Yuen (1993), and Melville and Gordon (1997) formed the main expansion of literature on defects.

A basic premise of this revised edition is that in order to understand best how buildings fail, one should first understand how they work. Construction technology is about how buildings work. Building Pathology is about how and why they fail. Thus, a sound knowledge of building construction is useful if not essential in understanding the causes and mechanisms of defects.

What makes this revised edition different from many of the other books on building problems is that it addresses the process of defect diagnosis in more detail. Chapter 1, therefore, includes an introduction to the nature and purpose of Building Pathology, the discipline relevant to the study of defects. This is followed by a new chapter on Building Diagnostics, which addresses the problem-solving and decision-making issues relating to investigating defects. Chapter 3 is a new section dealing with the basic methodology for investigating defects. Another new one is Chapter 4, which covers diagnostic aids such as chemical analysis and non-destructive testing that can be used to help achieve a more reliable diagnosis. The section on durability, now Chapter 5, includes coverage on service life assessment.

Also contained in this new edition are four new appendices containing a glossary and presenting several sample defects data schedules and reports.

Acknowledgements

Permissions

Improvements in building techniques, understanding of the reasons for, and the prevention of, failures, the encouragement of sensible innovation and the development of safe standards owe much to the Building Research Establishment (BRE), the now defunct Property Services Agency, the British Board of Agrément, the International Council for Research and Innovation in Building and Construction, and the British Standards Institution. So, too, does this book.

We also would like to thank the following organisations for their permission to reproduce these figures and tables in this book:

- Figures 3.3 and 3.4 are reproduced by the kind permission of the International Council for Research and Innovation in Building and Construction (CIB, its French acronym).
- Butterworth-Heinemann for permission to use Figure 3.6.
- Infrared Analysers Inc. for permission to use the photographs in Figures 4.10 and 4.11.
- British Standards Institution for agreeing to the publication of Figure 7.3 and Tables 4.1 and 4.2.
- Property Services Agency for permission to use Figures 7.2, 11.2, 13.1, 13.5, 13.6 and 14.1.
- Director of Building Research for Figures 5.1, 6.1, 6.2, 6.3, 6.4, 7.1, 7.4, 9.1, 9.4, 9.5, 9.6, 10.5, 10.6, 10.7, 10.8, 10.9, 10.10, 12.1, 12.2 and 14.2.
- Table 2.2 reprinted from *Fundamentals of Nursing* (5th Edition), P.A. Potter and A.G. Perry, Table 13.1, p. 275, Copyright 2001, with permission from Elsevier.
- Figures 9.2 and 10.3 are based on figures in publications by the Building Research Establishment.
- Figure 11.1 is reproduced by kind permission of the Brighton Evening News and Argos.

- Chartered Institution of Building Services Engineers for permission to include Table 5.3.
- Mitchell Beazley Publishers for the use of information resulting in Table 8.2.

Reviser's

First of all I would like to express my appreciation to Bill Ransom for allowing me to revise his book. It must be difficult for any author to hand over their work to someone else. Still, I hope Bill feels that the trust in this reviser was well placed.

Naturally, I would like to thank the publisher Taylor & Francis for asking me to undertake this revision. It is encouraging to see that an unglamorous subject such as building failures being given such support. Moreover, the construction industry needs to keep expanding its body of literature to enhance best practice on diagnosing defects and how to minimise if not avoid them.

My thanks in particular go to the commissioning editor Tony Moore for giving me the opportunity to revise this book. I also wish to express my appreciation to Dr Monika Faltejskova and Katy Low for their assistance and guidance throughout this revision.

Some of the insights on inspection methodology in this edition I owe to Daniel Friedman, a registered home inspector in the USA. His work in this area, particularly on how to detect, interpret and respond to difficult inspection conditions, was very helpful.

Wherever possible every attempt has been made to acknowledge the numerous sources used in this book. The author would like to apologise in advance if there is any case where this has not been achieved. Due acknowledgement will be made in any subsequent edition.

Finally, as ever, my thanks go to my family and colleagues for their patience and support during the revising of this book.

Disclaimer

Every effort has been made to check that each advice given in this book is correct and up-to-date. However, neither the author nor the publishers shall be held liable for any loss, damage or other negative consequence, no matter how they arise, occasioned by the implementation of such advice. Moreover, despite every effort to ensure accuracy, they cannot accept responsibility for incorrect or incomplete information, changes in standards, changes in product ingredients or availability, undiscovered hazards in materials and components, or any adverse health or other effects.

Chapter 1

Introduction to building pathology

OVERVIEW

This chapter outlines the principles of Building Pathology. It provides the historical and technological background for the systematic study of building problems.

1.1 Context

1.1.1 Problems of building failures

Most building defects are avoidable; they occur, in general, not through a lack of basic knowledge but by non-application or misapplication of it. Such knowledge seems to become mislaid from time to time. Those with long memories and those whose business it is to make a particular study of building defects are often struck by the re-emergence of problems which have been well researched and documented. Certain fundamental properties of materials, such as their ability to move through changes in temperature and moisture, seem to be overlooked and a rash of difficulties occurs. A call goes out for more research but, in truth, all that is usually needed is a good system for the retrieval of information, a better procedure for its dissemination and, most important, the realisation that a search for information is desirable.

Current training in design tends to concentrate on what to do rather than on what not to do. A similar situation exists in training in constructional techniques, where the craftsman is instructed how best to undertake a particular operation but to a lesser extent in the dangers of deviation from an accepted technique. Understanding of the likelihood of defects through inadequate design or construction is taught implicitly rather than explicitly in most built environment degree courses.

The level and nature of defects in building construction currently encountered suggest that more guidance is required on the avoidance of failures.

A need is seen, too, for such guidance to be a positive part of a training curriculum. Indeed there are good arguments for suggesting that, as the first essential in design and construction is to ensure that the structure provided is stable and durable and so specific education in the avoidance of failure should be a major part of any design and construction syllabus.

The purpose of this book is to provide such positive guidance in a suitably compressed form. It does not set out to describe every possible way in which a building may become defective: such a task would scarcely be possible and certainly would not be particularly helpful. It seems better to aim at identifying the principal defects and their causes, which if wholly eliminated would prevent the great majority of the defects which currently occur, save occupants of buildings much annoyance and discomfort, and also reduce the national bill on maintenance and repair by scores and, possibly, by hundreds, of millions of pounds annually.

The book aims to identify the nature and cause of important defects occurring in buildings, with emphasis on those affecting the fabric of a building and its associated services. It does not deal with issues of aesthetics, lighting, or thermal or acoustical comfort. While concerned primarily with the avoidance of defects, the text, particularly in Chapters 2, 3 and 4, also gives guidance to aid in their correct diagnosis when, unfortunately, the situation demands cure rather than prevention. Except in a general way, the repair of such defects is not covered. Any one specific failure needs a detailed examination to decide on the most appropriate repair, for this depends not only upon technical considerations but also upon the type of building and its age, and upon related economic and social considerations. There are few standard solutions to problems of repair.

Most defects occur through the effects of external agencies on building materials, and Chapters 5 and 6 consider in some detail the nature of these and their effects on the materials commonly used in building. These agencies include the principal components of the weather, namely solar radiation, moisture and air and its solid and gaseous contaminants; biological agencies, in particular fungi and insects; ground salts and waters; and manufactured products used in conjunction with building materials, for example, calcium chloride. Moisture occupies a central role, as the villain, in many building failures (Rose 2005).

The main sources of moisture and the ways in which the amounts present may be minimised are dealt with in Chapter 7. Special emphasis is given to the cause and effects of condensation, and how the risks may be avoided or reduced. Condensation, particularly in local-authority dwellings, can truly be said to have been the greatest single cause of human discomfort in dwellings since the end of the Second World War. The elements of building structure are then dealt with, starting with foundations and progressing logically upwards to roofs and parapets, passing on the way, floors, walls, cladding and external joinery. The avoidance of defects in building services has a chapter to itself. The book concludes with a more speculative chapter dealing with failure patterns and control. It attempts to relate defects to problems associated with the structure of the industry, to the dissemination of information and to particular difficulties which result from rapid innovation. Current control methods are outlined and a possible strategy is suggested for improving control, quality and reliability.

The intention and hope is that this book will provide positive guidance to the student designer, technologist and builder on how to diagnose and avoid the principal defects in buildings. It includes few complex scientific concepts and requires only a little special knowledge of science. Though concerned more with normal building than with major civil engineering construction, much of the text is of relevance to structural engineers also, particularly those parts dealing with the properties of the structural materials with foundations and with cladding.

The point was made at the beginning that knowledge gets mislaid: a further aim of this book is to serve as an aide-mémoire for practising designers and builders. For this reason, it has been kept concise, and is illustrated to give visual emphasis to some of the more important defects which can occur. These illustrations and parts of the text which describe the likely appearance of failures may assist surveyors and maintenance personnel, too, by steering them towards the probable cause of a failure. Though the essential aim is to avoid failure, once it has occurred and maintenance is needed, it is hoped the book will help both in identifying the cause and in preventing the adoption of the wrong remedial action. It may also help the maintenance engineer and surveyor by putting the severity of a failure and its consequences into a reasonable perspective and so prevent over-reaction to the event, which is not uncommon, particularly with foundation problems. If the book succeeds only partly in these ambitions it will, nevertheless, save both money and misery.

1.1.2 What is building pathology?

The scientific and technological discipline underpinning this edition is Building Pathology. It comes from the medical discipline Pathology, which in its simplest terms is the study of diseases. The word "pathology" is derived from two Latin words: *pathos* (disease) and *logos* (discourse).

According to CIB (1993), Building Pathology is defined as the systematic study or treatment of building defects, their causes (aetiology), their consequences and their remedies (or therapy). Harris (2001) stated that the term was introduced as a subject in its own right in an architectural programme at Columbia University by JM Fitch in the 1970s. Since that time, Building Pathology has gained increasing acceptance as an important part of architectural conservation and building technology.

Building Pathology involves a holistic approach to understand how the various mechanisms by which the material and environmental conditions within a building can be affected (Watt 1999). It is therefore comprehensive in scope, covering the investigative process from initial manifestation of the defect through to rectification and monitoring. Prevention, of course, is also a prime objective of Building Pathology, and is addressed in this book.

As highlighted by the International Council for Building (CIB 1993) the nature of Building Pathology can be described as follows:

In medical terms "pathology" has specific meanings as when:

- it refers to diseased conditions in relation to their determining causes (pathogenesis)
- it concerns macroscopic and microscopic alterations caused by such diseased conditions (anatomy and pathological hystology)
- it refers to general pictures of a disease (special pathologies), both medical and surgical.

Generally this term is related to illness; however, it may also be used to mean material and product alterations, extending also to treatment, prophylaxis and restoration procedures.

Such "metaphorical terminology" is useful in discussing this topic since the medical field is nowadays quite rightly concentrating on "prevention" rather than purely on the "treatment". Such an approach is also to be pursued in the building field.

"Pathology science" has recently become of major importance in building, strangely due to the consequence of innovation. In fact, research is quickly applied in practice, with the objective of shortening times which would, otherwise, be devoted to further experiments.

Not surprisingly, much of the procedures for investigating defects have emerged from diagnostics in the health-care professions. There are, however, major differences between medical and building diagnostics. First of all, medicine as a discipline can involve many diseases that are extremely complex and hard to diagnose early on (e.g. ataxia, multiple sclerosis, cancer, etc.). Some ailments may be so difficult to diagnose that they are termed "idiopathic". This is slightly mitigated by the fact that health care deals with animated objects – human beings – which are directly responsive to stimuli. Patients can respond to questions and give the clinician direct feedback on what is ailing them. Sophisticated medical diagnostic tests on previously inaccessible parts of the human body can be easily carried out nowadays (e.g. CAT/CT [computerised axial tomography] scan, MRI [magnetic resonance imaging] scan, micro fibre-optic arterial inspections, etc.). The initial patient information from such tests is often an essential part of a clinician's diagnostic investigation (Barrows and Pickell 1991).

In contrast, buildings are inanimate objects and therefore cannot give direct feedback on problems they have. The building pathologist thus has to rely on indirect feedback in the form of evidence, clues or signs and symptoms, all of which should lead to inferences that can aid diagnosis.

Building Pathology is also related to forensic engineering. The latter is the specialist discipline within forensic science that investigates for legal reasons the causes of failures of devices and structures. Thus, according to Kaminetzky (1992) there are three essential elements in forensic engineering's investigative circle: (i) act with speed to collect data, (ii) determine failure pattern and (iii) establish correlation between analysis and field measurements. Chapter 2 shows that these elements are also applicable to Building Pathology.

The Latham (1994) and Egan (1998) reports highlighted the need for the construction industry to increase its efficiency and suggested ways to achieve this. These led to the formulation of key construction performance criteria. It is now generally accepted that a building should be constructed/adapted on time/schedule, within budget, without any major accidents or injuries, and the finished product should exceed client's expectations and have zero defects. In view of the last mentioned criterion, Building Pathology has inevitably become more important.

Another driver in the United Kingdom for increasing the importance of Building Pathology is the Home Condition Report (HCR). A major component of the HCR is the assessment of defects (Parnham and Rispin 2001). This along with the Energy Performance Certificate (EPC) forms part of the Home Information Pack (HIP), which is designed to improve the process of buying and selling a home. However, the Department of Communities and Local Government (DCLG) announced in July 2006 that there would be a phased roll-out of HIPs in June 2007, with only EPCs and not HCRs being compulsory initially.

Two other examples of this increase in the profile of Building Pathology relate to changes made by two professional construction bodies to their prescribed core competencies. In the late 1990s, both the Royal Institution of Chartered Surveyors (RICS) and the Association of Building Engineers (ABE) expressly incorporated Building Pathology in their list of competences as a core area of skill and knowledge for building surveyors. As a result, it is now a key subject area in the advanced stages of most Building Surveying degree courses.

Building pathologists, however, may come from a variety of disciplines in the built environment. Chartered building surveyors, of course, make natural building pathologists. Other cognate professionals, such as architects, engineers and construction managers, with years of experience dealing with defects investigation are also suitably qualified to undertake this specialist role.

1.1.3   W086 objectives

The CIB has over 50 working commissions on a whole range of issues relating to construction and buildings. One of these commissions is W086: Building Pathology (CIB, 2003a). It was founded in 1985 (Freeman 1987). The objectives of W086 are

- To produce information that will assist in the effective management of service loss.
- To develop methodologies for assessment of defects and failures and consequential service loss.
- To apply systematic approaches to the investigation and diagnosis of defects and failures in buildings of all types and at all stages of life.
- To audit buildings in use to check the veracity of service loss prediction.
- To promulgate findings to all those involved in the production and management of buildings.

This edition aims to contribute in particular to the second and third criteria listed above.

1.1.4   Distinction between defect and failure

At this juncture it would be useful to clarify the distinction between a building defect and a building failure (see also Appendix A: Glossary, which gives definitions of these and other Building Pathology-related terms). Strictly speaking, these two terms, although similar, should not be treated as being synonymous. A defect is a shortfall in performance occurring at any time in the life of the product, element or building in which it occurs (BRE Digest 268). It is also a departure from design requirements where these were not themselves at fault. A failure, in contrast, is the termination of a product or element's ability to perform its intended function.

For the purpose of these notes, then, a defect shall be interpreted as a fault in an element, material or component of a building. For example, a flush cope on a parapet with little or no flashing would be termed a defect. On the other hand, a building failure is here meant as a consequence of such a defect/fault. A failure may be thought of as an unplanned or unintentional negative effect of one or a combination of faults, which leads to a shortfall in performance. Using the flush cope example given above, a typical failure would be rainwater penetration at the wallhead due to the inadequate weathering of the cope, which results in spalling or lamination of the brickwork/rendering. It is this spalling that would be deemed the failure using the foregoing definition.

Moreover, in law a distinction would be made between a fault and a failure. The former would be classified as a defect, that which caused or triggered the failure. The latter would be classed as the damage resulting from such a defect. This is important for the courts when it comes to assessing the cost of remedying construction failures to ascertain the extent of damages awarded to the aggrieved party if a case has been proved against the contractor or designer.

The main emphasis of this book will be on functional or technical failures of the non-structural kind. It will also give some consideration to basic structural failures, but failures in aesthetics, amenity or planning will not be addressed.

According to Puller-Strecker (1990),

> Defects then can be real or perceived. Even if they are real, they are relative rather than absolute. Their origins are often not what they seem to be, but can be traced back to a complicated interaction between lack of knowledge, lack of education, lack of training, lack of experience, lack of definition, lack of care, or lack of enforcement. Because they are relative, defects cannot be avoided, but by the same argument, quality can always be improved and defectiveness can always be reduced . . .
>
> . . . A defect may be described as something that:

- does not come up to the expectations of the client;
- falls below the prescribed standard for things of its kind;
- is less acceptable than it ought to be, bearing in mind the state of the art and economic reality, or
- is the result of an error.

In this day and age, one could reasonably assume that since the causes of most defects are fairly well known there should be an overall decline in building failures. Unfortunately, the reverse is the case. Some have argued that, if anything, building failures are increasing in frequency and severity (CIB, 2003b).

Reducing if not eliminating the incidence of building failures is not just important to curb construction costs. Clients are more demanding and their expectations of better construction standards are increasing. "Lean construction", with its emphasis on waste minimisation, supply chain management and fast track construction, is just one of the many initiatives responding to these pressures.

Moreover, we need to control building failures for the following main reasons:

- Increase the sustainability of buildings – make them last longer and more energy efficient.
- Achieve buildings that offer better value for money with lower maintenance costs.

- Make properties safer – reduce hazards and accidents in buildings.
- Combat poor indoor air quality – to avoid sick-building syndrome and building-related illnesses.
- Minimise negligence and other legal consequences in construction.

1.1.5 Taxonomy of failures

Failures of building products, components and elements come in all forms and levels of severity. The type of failure will impact on both the level of

Table 1.1 Typical categories of failures

Failure type	Example
Aesthetic failure	• Crazing or shrinkage cracking of concrete or render. • Flaking and peeling of paintwork. • Bossing and spalling of render. • Staining and soiling of finishes. • Chipped, dented or lipped floor/wall/ceiling finishes and veneer finishes to doors.
Functional failure	• Misalignment of building components such as doors and windows not operating properly. • Leaks in elements such as roofs, walls and floors. • Sagging of floors.
Failure of materials	• Chemical attack of rendering, mortar or brick. • Fungal attack of timber. • Corrosion of metals.
System failure of components and elements	• Carbonation of concrete, leading to corrosion of reinforcement and subsequent cracking and spalling of concrete members. • Debonding and bubbling of membrane from substrate owing to moisture or incompatibility.
Structural failure	• Subsidence (a downward movement of a building caused by below ground factors – such as desiccation of clay soil) • Settlement (a downward movement of a building caused by above ground factors – such as overloading).
Non-structural failure	• Delamination of roof tiles and slates. • Cracking and debonding of plaster or rendering. • Blistering and peeling of paint coatings. • Tenting, debonding and bubbling of floor coverings.
Reversible failure	• Jamming of doors and windows as a result of moisture intake by these components – usually in winter; in the summer the wood dries out and the windows and doors become unstuck.
Irreversible failure	• Chemical reactions such as sulphate attack on mortar or rendering. • Excessive distortion in beam/slab, column or wall owing to structural movement

investigation and the required therapy. (See BRE Digest 268 and IP 15/90 for data on failure patterns associated with housing.)

Building failures can be classified into several broad groups as listed in Table 1.1. These failure classifications, however, are not mutually exclusive – some overlap with one another. For example, the failure of a material can be both irreversible and non-structural – such as bossing and spalling of plasterwork/rendering.

Building failures can be categorised into a variety of groups depending on their nature and source. They can be classed as physical (structural) failures (which result in loss of certain characteristics) or performance failures (which mean a reduction in function below an established acceptable limit). Heckroodt (2002), on the other hand, uses the terms "system failure" and "material failure". These latter categories are summarised in Table 1.2.

Note that there is often an overlap between the failure categories listed. For example, a structural failure can lead to a non-structural failure

Table 1.2 Other categories of failure

Categories of failures		Examples
Physical	Non-Structural	• Bossed or spalled rendering. • Flaking and peeling paint. • Dampness. • Disintegration of wall finishes. • Softening or deterioration of insulation.
	Structural	• Cracking in wall, floor, column or beam. • Buckled or twisted column, tie, strut or beam. • Deflected or sagged floor structure. • Subsidence. • Leaning or bowing wall.
Contextual	System inter-component	• Debonding of floor/roof covering. • Blocked outlet clogging gutter. • Wall tie corrosion leading to cracking and distortion of brickwork. • Failure of sealing gasket in double glazing unit. • Masonry mortar too weak/strong.
	Material intra-component	• Corrosion of metalwork. • Sulphate attack on concrete and mortar. • Frost action in brickwork. • Fungal attack in wood. • Insect infestation of wood.
Observational	Conspicuous	• Failure to detect obvious defect. • Ignoring significance of minor obvious defect.
	Hidden	• Failure to detect inconspicuous defect • Ignoring symptoms or implications of possible hidden defect.

(e.g. subsidence can lead to cracking in rendering) and vice versa (e.g. a dampness problem can lead to fungal attack of loadbearing timbers nearby). A material failure can lead to a system failure (e.g. corroded wall ties can result in cracking and bowing of outer leaf of brickwork) and vice versa.

1.2 Background to the construction industry

1.2.1 Failures in construction

There are a number of reasons why building professionals should study the nature and consequences of defects and failures. First, the cost of repairs and corrective maintenance runs into many millions of pounds each year (Houghton-Evans 2005). As a result, scarce resources are expended on rectifying many defects that need not have occurred in the first place. Our aim should be to avoid or at least minimise the extent of such building problems.

Secondly, the presence and proliferation of faults, particularly within housing, causes needless annoyance or distress to the occupants. The Latham and Egan reports (1994 and 1998 respectively) emphasised the importance of considering the client's needs and expectations.

Thirdly, the reputation of the industry is being undermined by a seemingly endless stream of incompetent, embarrassing and sometimes dangerous series of failures in construction work. This is reflected in the number of litigation cases against building professionals and contractors.

Every type of building contains defects to some degree. The consequences of some defects may be minor but others are more important and may affect the appearance and usage of the building. However, the cost of rectifying or (worse still) not rectifying building defects could be extensive. In more serious instances, they may pose a hazard to health and safety.

As stated in the beginning of this chapter the majority of building defects are preventable: their reoccurrence is more usually caused by non-application of basic knowledge rather than by a lack of such knowledge. This implies that it is not necessarily the expertise that is at fault; it is rather the attitude to the job of those in the industry, which is sometimes lax. This lack of care is the underlying cause of many building problems.

The initial cost of most buildings is high because of the labour-intensive, ad-hoc nature and scale of construction works. As already indicated, it is even more costly and awkward to rectify a failure than it is to prevent it from occurring in the first place. For instance, the installation of movement joints in the external walls of a building is a fairly straightforward operation during its construction. If such joints had been omitted at that stage, installing them at a later date would be troublesome and expensive for both the builder and occupier. Imagine the cost of hiring operatives to carry out this work; materials for the job; the difficulty in obtaining proper access; and the inconvenience and disruption to the occupants.

Clearly, therefore, it is easier and less costly to "get it right" initially than to "put it right" afterwards. As was indicated earlier, some building failures can be difficult and expensive to rectify completely, if at all. In a Building Research Establishment (BRE) report (IP 15/90), it was reckoned that it may cost five times as much to rectify a defect than to get it right in the first place.

1.2.2 Characteristics of the construction industry

To understand why defects and failures seem such an intractable problem one must appreciate the peculiar nature of the building industry. Construction is one of the few major industries in this country where generally there is a separation in the design and manufacturing functions. The nearest equivalent to the car manufacturing industry in construction is the "design and build" contract in which the contractor provides both the design and construction services for a particular project.

Another distinctive feature of the construction industry is that, unlike most other goods or products produced by other industries, each building project is different or unique in some respect. There are a number of factors that give construction projects their special characteristics; a few of these are listed below:

- Variable site conditions. No two building plots are identical in terms of topography, access, subsoil, climate/exposure, available services, etc.
- Restraints such as planning control and building regulations vary in implementation from region to region, and these change over periods of time.
- Most building projects take many months if not more to complete. This time factor makes many sites susceptible to disruption and inclement weather.
- The people involved in each building project may change from month to month. No two construction jobs have the same set of designers, advisors, supervisors or operatives. This invariably leads to variations in quality. There is, moreover, a certain lack of continuity in construction jobs often resulting from a change in the workforce composition.
- Innovative materials or components or techniques or old materials used in new situations are continually being tried out in building without much knowledge of their long-term performance features.
- Much of the work is carried out in the open. Thus, a great deal of the work is not only exposed to the elements but is also carried out in non-factory conditions. This makes it harder to maintain a stable environment to undertake work undisturbed by wind and rain.

- Materials and components do not necessarily come from the same sources for all contracts. This tends to lead to greater variations in quality.
- Because of the dichotomy between the designer and contractor indicated above, architects do not normally tell builders how to erect buildings. In other words, the client, via the designer, tells the builder what is required – he does not tell the contractor how to go about it, unless special site or contractual considerations are present. This may lead to some changes in the construction during the contract by the builder to suit his programme.
- Quality control is not consistent from site to site. For instance, standards of supervision by site agents and clerks of works will depend upon the respective backgrounds of those involved, their attention to detail and the time spent on site, etc.
- Feedback between the user or installer and designer is often slow or difficult if not non-existent. The cause of this problem was highlighted by an engineer, Pullar-Strecker (1990), in a CIB keynote address:

> In the manufacturing industries, the feedback of performance data is one of the main sources for improved design, and in the consumer market where the life of products is short and turnover fast, the pace of improvement that results from this process is commercially vital because it fuels the market that can be generated by planned obsolescence. In the construction industry, in the UK at least, a similar system of feedback and improvement does not seem to exist, nor do we even have a basis on which one could be constructed.

There are a number of reasons for this state of affairs. Buildings, in comparison with manufactured goods, have a relatively long working life, often much longer than the working life of those who designed or built them, and reliability of performance or durability of fabric are judged on timescales which may span several generation of designers. It is inevitable therefore that designers cannot follow through the performance of their own buildings for very long, but in practice most designers lose touch with their buildings much more quickly than this and seldom have any real opportunity to study how well they perform, even initially. This factor taken together with the fact that, unlike manufacturing industry, the building industry sells its prototypes but usually produces no large numbers of "production" models means that individual designers must learn from the performance of buildings designed and built by others rather than themselves.

It is clear therefore that building defects and failures account for a large proportion of expenditure in construction. There are two aspects of the industry that have contributed to this problem: its fragmentation and the overemphasis of education and training on new-build.

(i) The fragmentation of the construction industry.

Unlike car manufacturing, for example, the building industry is very fragmented. It contains many thousands of small to medium size firms rather than only several large companies. This inevitably leads to fewer economies of scale and to reduced levels of cross-information and feedback. The repair and maintenance sector is even more diverse than the new-build part of the construction industry.

Another aspect of this problem is the fragmentation of the building professions within the United Kingdom. One of the distinctive features of British construction practice is the tendency for overspecialisation to occur early in the training of construction professionals: architects; quantity surveyors; structural engineers; services engineers; construction managers; building surveyors; valuation surveyors; and so on. In the United States and many European countries on the other hand there are only two principal professional groups in construction whose status is legally protected: architects and engineers. There are specialists within each of these two disciplines that who carry out all of the functions normally fulfilled by construction surveyors/technologists or construction managers/economists in this country.

(ii) An overemphasis on new-build in education and training.

Despite the fact that it accounts for just under half of the total output of the construction industry, the training of most building professionals, managers and operatives is on new-build work. In contrast, there seems to be little academic or professional interest in the so-called "Cinderella" subject of maintenance/repair and refurbishment. This prejudice and lack of appreciation will take years to overcome.

Still, many construction professionals are realising that problems in building maintenance are just as complex, if not more so, than those in new-build. For example, the late Sir Ian Dixon, president of the CIOB in 1990, stated in another CIB keynote address that

> The technology of building maintenance is sometimes characterised as being at the bottom end of a scale of industrial sophistication. This is completely inaccurate. Maintenance deals with the building as it is, not as the contractor might wish it to be. Indeed, it is my experience that the technological problems encountered by much building maintenance work is far more demanding than that of a new building site, and the need for extensive experience is paramount especially when dealing with dampness problems, whether they come from the top, bottom, inside or outside of a building . . .

1.2.3 Human sources of building defects

Building defects continue to occur because of economic pressures to reduce costs or cut corners, and reliance on untested innovation coupled with poor communication and a failure to adhere to the available recommendations for good building practice (Addleson, 1992; Houghton-Evans 2005). All of these influences can trigger the following main categories of errors (CIB 1993):

- Pre-design errors: e.g. poor or inadequate brief.
- Design errors: e.g. unusual or awkward details leading to leaks.
- Construction errors: e.g. bad workmanship.
- Maintenance errors: e.g. lack of regular cleaning out of roof gutters.

Research carried out over the last thirty years by the BRE (IP 15/90) and others such as CIB (1993) has suggested that building defects in percentage of main origins are distributed approximately along the lines indicated in Table 1.3 below. Inadequate maintenance as well as user misuse and abuse, of course, will exacerbate such problems.

1.2.4 Typical problem areas

Many construction failures can be attributed to inadequate communication, lack of awareness of the need for greater acceptance of responsibility on the part of all those involved in the construction process, slack controls over building works, and ill-considered introductions of innovative materials and techniques.

(a) Communication breakdowns
These account for a whole range of problems in society, even in this so-called information technology age. Communication is important at all levels

Table 1.3 Distribution of human sources of construction defects

Sources	UK (%)	Ave. for rest of the developed world (%)
Construction: • Poor workmanship • Inadequate Supervision • Vandalism	50	40
Design: • Poor detailing • Inappropriate specification • Inferior quality of design or materials used	40	45
Products: • Faulty manufacture • Damage as a result of faulty delivery • Damage resulting from inadequate storage or protection	10	15

in the building process, from the client to the designer, and designer to builder, and so on. Clients must convey to their professional advisors their requirements for their intended projects in terms of type, style and size of building, available budget, timescale involved, etc. In turn, the architect and other members of the design team (which may include a building/maintenance surveyor) must translate these needs to the builder in as clear and as unambiguous terms as possible. Clients' requirements take the form of written briefs, which will be converted into feasibility studies by professional advisors. If approved by the client, the design team will then convert the outline proposals into layout plans, detailed drawings, specifications, bills of quantities, etc. In addition, other instructions either oral or in writing will normally need to be given during the contract period to cater for contingencies and clarify obscure or unclear instructions, which inevitably occur in construction projects.

Nowadays there is a changing attitude to building design and technology. A typical example of this shift in emphasis is the increasing use of the performance approach to building. This approach is above all the practice of working in terms of ends rather than means. The former is concerned with the functional requirements of a building or building product, with identifying the end result. The latter, on the other hand, which is the more traditional approach, is concerned with how a building or product is to be constructed, with prescribing the recipe for the work. With the performance approach, good communication is also essential. Designers must be able to gather data from different sources and formulate those data into a performance specification that is articulate and clear.

Another inherent problem within the construction industry is that many parts of it are run on an adversarial or confrontational basis. This type of negative relationship, which is sometimes engendered between clients' representatives and contractors, is not conducive to making building projects run smoothly or efficiently.

Poor communication can occur in a number of ways: designers/specifiers being unaware of the requirements in the codes, standards, regulations or elsewhere; professionals assuming incorrectly on the evidence of the subsequent failure that site expertise could be relied upon; inadequate information supplied in the drawings/specifications; etc.

(b) Lack of responsibility and accountability

All participants in the building process, from designers to site managers through to operatives, have an ethical as well as a contractual responsibility to carry out their jobs to the best of their ability. This is not meant to sound patronising or sanctimonious, but indicates one of the root human causes of building defects: cavalier, indifferent or negative attitudes towards implementing building work by some designers, managers and operatives. Naturally, the solution to this problem lies with society as a whole rather than with just those in the construction industry.

On site, for instance, the majority of defects tend to emanate from lack of care rather than a lack of skill. Shortcuts in and disregard for good standards of workmanship, and general carelessness on site arise for a number of reasons:

- Cost cutting leading to difficult or poor working conditions on site.
- Unrealistic deadlines or bonus work encouraging workers to rush their jobs, which often leads to mistakes or skimping on standards.
- Indifference or lack of ethical commitment on the part of some supervisors and operatives.
- No incentives to work at best performance owing to poor basic pay and lack of differentials for skills.

(c) Inadequate controls

To ensure that a building project is going according to plan, some forms of control must be exercised by the client's representatives as well as by the contractor's agents at all stages of the construction work. Building failures can occur owing to a lack of systematic checking of the work during and at the end of the design stage as well as regular intervals on the job. In particular, site control and supervision by the contractor's representative, normally the site agent, and the client's/architect's agent, normally the clerk of works, is crucial to the success of any building project. Such controls could be made more stringent, and hence more effective, in a variety of ways:

- Improve the quality and quantity of on-site supervision. For example, clerk of works and site agents could be given more training in building technology and better man-management.
- Arrange for more periodic, unannounced, comprehensive on-site inspections.
- Clerk of works could "rotate" between sites during the construction of several projects.
- More emphasis and encouragement should be given to engender or instil common-goal attitudes amongst operatives and supervisors.

(d) Innovative materials and techniques

There is of course nothing inherently wrong with introducing new products and processes in construction. The main requirement for doing so is that they are adequately tried and tested, and those installing them are aware of any special fixing requirements. This issue is addressed further in Chapter 15.

1.3   A model of best practice in building pathology

The model adopted by the reviser of this book contains three pillars as shown in Figure 1.1. The ethical considerations that underpin this model

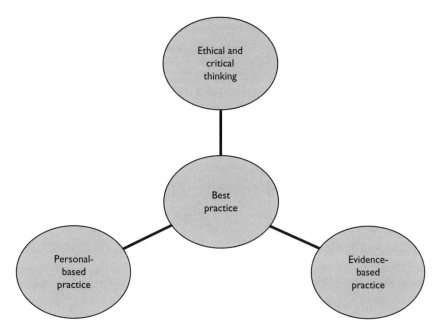

Figure 1.1 The three pillars of best practice in Building Pathology.

and the critical thinking addressed in Chapter 2 are encapsulated in the nine core values of the Royal Institution of Chartered Surveyors (RICS), one of the construction industry's premier professional bodies:

1. *Act with integrity*: Never put your own gain above the welfare of your clients, and respect their confidentiality at all times.
2. *Always be honest*: Be trustworthy in all that you do, never deliberately mislead, whether by withholding or distorting information.
3. *Be open and transparent in your dealings*: Share the full facts with your clients, making things as plain and intelligible as possible.
4. *Be accountable for all your actions*: Never commit to more than you can deliver, take full responsibility, and don't blame others if things go wrong.
5. *Know and act within your limitations*: Be aware of the limits of your competence and don't be tempted to work beyond these.
6. *Be objective at all times*: Give fair neutral advice, and never let your own feelings or interests cloud your judgement.
7. *Never discriminate against others*: Always treat others with respect whatever their gender, race, religion or sexual orientation.

8. *Set a good example*: Remember that both your public and private behaviour could affect your own reputation, RICS' and other members' reputations.
9. *Have the courage to make a stand*: Be prepared to act if you suspect another member of malpractice.

1.4 Evidence-based practice

1.4.1 Definition

Effective diagnosis is best achieved and validated if it is evidence based. Evidence-based practice (EBP) is already well established in the health-care professions (Sackett *et al.* 1996). To paraphrase the standard medical-orientated definition for its application to Building Pathology, EBP can be described as the conscientious, explicit and judicious use of current best evidence in making decisions about the diagnosis and treatment of defects in buildings.

The main elements of EBP are

- Scientific evidence (from on site and laboratory analysis)
- Client references or values
- Context – building/client circumstances
- Practitioner experience and judgment.

Evidence is the available body of facts, information and data that people select to help them answer questions. It may be direct (i.e. high moisture content readings in the lower part of a wall indicating, if not confirming, a dampness problem) or indirect (i.e. tide mark on a wall above skirting indicative of a dampness problem).

The following are other classifications of evidence:

- *Primary/Secondary evidence*: Primary evidence is first-hand reports/feedback/data on a defect. For example, it can comprise moisture readings using a "Protimeter" or calcium carbide meter. Secondary evidence is reports and feedback from others (e.g. occupier) on the defect.
- *Witting/Unwitting evidence.* Witting evidence is the documentary information intentionally imparted by the original author of a relevant document or persons involved with the building under investigation. Unwitting evidence is anything else that can be learned from relevant documents (e.g. drawings, reports, etc.) or building (Bell 1999).

Unfortunately, however, evidence is not value free. It is interpretation is context dependent and it is often incomplete or equivocal. Where it is available, the evidence may not be fit for purpose – i.e. its quality may be poor. Moreover, evidence can be complex, difficult to interpret and open to more than one interpretation. Evidence, therefore, needs to be mediated, and getting it is often costly, slow and difficult.

The following criteria can be used to assess evidence in a building investigation:

- Measurable/recordable?
- Verifiable?
- Pattern of appearance (chronologically or locationally)?
- Form of appearance?
- Position?
- Element/s affected?
- Suspects – hypothesised?
- Aim in dampness investigation: e.g. Identify/confirm moisture source?

1.4.2   Requirements for implementing EBP

In order for the building pathologist to undertake EBP, sufficient research must have been published on the specific topic. This is clearly a limiting factor for construction practitioners. Unlike the health-care sector, which has hundreds of journals and other research material, the construction industry has only a dozen or so such publications. The main UK-based technical journals in construction dealing with defects are *Structural Survey (The Journal of Building Pathology & Refurbishment)* and the *Journal of Building Appraisal*.

Second, the building pathologist's practice must allow him/her to implement changes based on EBP. This may prove difficult where recognition and use of EBP is lacking.

Third, the building pathologist must have skill in accessing and critically analysing research. This can be achieved using the following four-step problem-solving approach:

1. Clearly identify the issue or problem based on accurate analysis of current construction knowledge and practice.
2. Search the literature for relevant material.
3. Evaluate the research evidence using established criteria regarding scientific merit.
4. Choose interventions and justify the selection with the most valid evidence.

1.4.3   Sources for EBP

How one uses EBP in Building Pathology has not, however, been as well developed in construction as it has in the health-care sector. Still, the main sources for data and other evidence on building defects and failures are as follows:

(a) Research bodies:

- Building Research Establishment: Digest, Reports, Good Repair Guides, etc.
- Construction Industry Research and Information Association (CIRIA): Reports and guides.
- International Council for Research and Innovation in Building and Construction (CIB): WO86 (Building Pathology).

(b) House condition surveys:

- House Condition Surveys Team (2004) *Scottish House Condition Survey 2002* (Main Report), Communities Scotland, Edinburgh.
- Research, Analysis and Evaluation Division (2003) *English House Condition Survey 2001* (Main Report), Office of the Deputy Prime Minister (ODPM), London.

(c) Building standards:

- Approved Documents (robust details): DCLG website
- Technical Standards (robust details): Scottish Building Standards Agency website.

(d) Textbooks:

- Addleson (1992)
- Carillion Services (2001)
- Hinks and Cook (1997)
- Marshall *et al* (2003)
- NBA (1983)
- Harris (2001)
- Richardson (2001)
- Watt (1999).

1.4.4 Problems with implementing EBP

EBP has more than its fair share of critics in the health-care professions. In summary, the main perceived limitations of EBP are as follows:

- It might undermine personal-based practice (PBP)
- Promotes a "cookbook" approach to decision making
- Ignores clients' values and preferences
- Is limited to scientific research
- May be perceived as a cost-cutting tool.

As with any initiative there may be some resistance to the full implementation of EBP. The main barriers to implementing EBP are

- FACTOR 1: Accessibility of research findings is often difficult.
- FACTOR 2: Anticipated outcomes of using research may not be recognised.
- FACTOR 3: Organisational support to use research may be minimal.
- FACTOR 4: Support from others to use research may not be present.

1.5 Personal-based practice

1.5.1 Background

To use a metaphor, this is the other wing of the best-practice bird. Without personal-based practice (PBP) the use of EBP would be ineffective as the data and other evidence collected needs to be contextualised and evaluated in a rigorous and systematic way as demonstrated below.

Personal-based practice is taken by the author to mean the consistent, explicit and prudent use of one's own proven abilities, experience and expertise, and *critically* reflecting on them in solving problems and making decisions. Critical reflection is necessary to minimise excessive bias and prejudice, both of which can result in a misdiagnosis.

Reflective practice, of course, is neither common nor explicit in construction disciplines. Thanks to the work of pioneers such as Schon (1995), however, it is now generally recognised as an important part of professional as well as educational practice. It is increasingly seen as a key factor in underpinning learning. It also enriches learning dialogues by getting novices (students and trainees) and practitioners to think more systematically about what they are doing (Race 2004).

1.5.2 What is reflection?

Reflection is a process of thinking back or recalling an event to discover the meaning and purpose of that event (Moon 1999). It is "a critical thinking process (see Chapter 2) that causes us to make sense of what we have learned, why we learned it and how that particular increment of learning took place". In other words, reflection is "a complex and deliberate process of thinking about and interpreting experience in order to learn from it" (Wilkinson 1999).

Reflective practice is now a key learning tool used throughout many higher education courses and beyond. It is a way of ensuring ongoing scrutiny and improved substantive practice across a variety of educational and professional settings. It provides a framework for organising one's thinking (Schon 1995).

Reflection allows a person to think about the things they have seen and done and to consider whether they were satisfied with their performance or whether they would handle the situation differently in the future. It is a form of "metacognition", a higher form of cognitive process – sometimes referred to as "thinking about thinking" or "speculative thinking" or "critical consciousness" (O'Callahan 2005). For example, one could think about

- how the situation helps you with your personal learning;
- the practical or professional skills you used; and
- the interpersonal skills you used or might have used.

1.5.3 Types of reflection

Reflection can be categorised into five main types, each of which can be used in defects diagnosis depending on the circumstances:

1. **Analytical reflection** or *process analysis* such as thinking about how one did something.
2. **Evaluative reflection** or *judgment assessment* such as assessing how well/badly one did something or how well/badly an event went.
3. **Speculative reflection** or *open-ended reflection* that addresses worrying or troublesome questions.
4. **Serendipitous reflection** or *unplanned reflection* that can still be useful with the possibility that "habits of questioning and reflecting developed in the other forms of facilitated reflection will carry over to good effect".
5. **Critical incident reflection** such as the Atkins and Murphy (1994) model used in the nursing profession (see below).

1.5.4 Stages of reflection

Reflection can occur at any time relating to an investigation. There are, however, three main stages of reflection:

1. **Reflection *for* action** – thinking proactively about what is the problem and how it could be tackled (McAleese 2003). It could involve, for example, asking questions such as, What is my prior knowledge that could help me solve this particular defect? How much time do I have to complete the task?
2. **Reflection *in* action** – thinking about developments as one goes through the process and identifying options for solving the problem (Cowan 1998). Questions that would be appropriate at this stage are, How is this investigation going? What do I need to do if I do not understand the defect?

3. **Reflection *on* action** – thinking about how effective the initial diagnosis was or how well the procedure was undertaken (Cowan 1998). Typical questions that could be asked at this stage are, How well did I do the investigation? What could I have done differently?

1.5.5 Why should one reflect?

Self-reflection has many benefits – not least of which is that it provides the opportunity to analyse one's practice. The following are the main reasons for reflecting on a problem or situation or event:

- It enhances learning.
- It helps us to gain insights into failures.
- It helps integrate theory and practice.
- It increases self-awareness and insight into behaviour.
- It allows critical analysis of practice.

According to one educationalist (Race 2004),

Reflection deepens learning. The act of reflecting is one that causes us to make sense of what we've learned, why we learned it, and how that particular increment of learning took place. Moreover, reflection is about linking one increment of learning to the wider perspective of learning – heading towards seeing the bigger picture. Reflection is equally useful when our learning has been unsuccessful – in such cases indeed reflection can often give us insights into what may have gone wrong with our learning, and how on a future occasion we might avoid now-known pitfalls. Most of all, however, it is increasingly recognised that reflection is an important transferable skill, and is much valued by all around us, in employment, as well as in life in general. The ability to reflect is one of the most advanced manifestations of owning – and being in control of – a human brain. Have you reflected today? Almost certainly "yes!". But have you evidenced your reflection today? Almost certainly "sorry, too busy at the moment". And the danger remains that even the best of reflection is volatile – it evaporates away unless we stop in our tracks to make one or other kind of crystallisation of it – some evidence. In our busy professional lives, we rarely make the time available to evidence our ongoing reflection. But we're already into an era where our higher education systems are beginning to not only encourage, but also to require our students to evidence their reflection. So what can we do to address the reflection culture gap – how can we approach accommodating our lack of experience in evidencing our reflection, and helping our students to gain their skills at evidencing their reflection?

1.5.6 How should one reflect?

Reflection is a cognitive process that enables the practitioner to explore, understand and develop meaning. It also helps highlight contradictions between theory and practice (O'Callahan 2005). The way one can reflect is to

- Think of a critical incident or process that you experienced (see the reflective cycle mentioned below).
- Ask yourself a number of key questions:
  - How did I do that?
  - How well did I feel I performed the task?
  - Any uncomfortable thoughts or feelings?
  - What was important, meaningful, good and decisive about my problem solving and learning? For example, how will you know what is meaningful? You might decide that a meaningful event is one that (a) you think about after it has occurred; and (b) that makes you associated something you already know with the event in question.
  - Are there any lessons to be learned from that experience?

All of these questions would be asked in a good reflective model. The use of a reflective model helps to structure one's thoughts and makes one analyse one's practice. A popular example used in the health-care professions is the model by Atkins and Murphy (1994), which involves:

- addressing uncomfortable thoughts and feelings
- describing the situation, including thoughts and feelings
- analysing thoughts and feelings relevant to the situation to identify knowledge, challenge assumptions, and imagine and explore alternatives
- evaluating the relevance of the knowledge to see if it helps to explain/ solve problems and to determine how complete was one's knowledge
- identifying any new learning that has occurred
- implementing action or new experience
- repeating cycle if necessary.

Such a reflective model can trigger the following three basic questions, which are useful for enhancing learning:

- What was the original gap in my knowledge of the defect/subject/issue investigated?
- What learning did I achieve in the course of that investigation?
- What are my further learning needs in relation to this defect/subject/ issue?

1.5.7 Praxis

On its own, of course, reflecting on practice will not result in solving a building problem. Action is required to implement the results of such metacognition. This conjoining of "thinking" and "doing" is called "praxis" (Lutz *et al.* 1997). It is a process that is underpinned by knowledge and action (O'Callahan 2005).

Thus, a building pathologist can use expert practice to develop reflection and expand one's thinking. As a result, it should heighten one's awareness of how one thinks and acts. For example, this could prompt a reflective account of the investigation into a suspected condensation problem. It can expand the investigator's awareness of the need to critically evaluate the evidence gathered – such as hygrothermal readings – to determine whether or not they support the investigator's initial hypothesis. In so doing, the building pathologist can turn action and thinking to strategic use and learn through practising.

1.5.8 Criticisms of PBP

Like any other approach, PBP is not without its weaknesses, as summarised below:

- Too much reflection could encourage excessive introspection or "navel gazing" or inhibit creativity (Day 2003). This may engender excessive doubt or scepticism, which in turn might lead to a paralysis in decision making (see "analysis paralysis" problem in the next chapter). It may also impede the "blink" or intuitive faculty that experienced practitioners employ to help them make effective rapid decisions (Gladwell, 2005).
- The extent and style of PBP is limited or particular to each practitioner. Not all practitioners will apply PBP to the same extent or with the same consistency.
- PBP might discourage adopting best practice from other sources if they conflict with it.

1.6 Summary

The United Kingdom's construction industry has set itself an ambitious target of zero defects. All the indications from CIB (2003b) and other authoritative sources suggest that building defects are not going to go away completely. Human imperfections, client pressures and unforeseen circumstances will continue to inhibit the achievement of new and adapted buildings that contain no defects. But this is no reason to prevent the industry and its professionals from striving towards best practice. The result will be better and more functional buildings that give clients and occupiers best value for money and provide society with durable and efficient constructed assets (BS 7543 and ISO 15686-1).

According to one writer (Puller-Strecker 1990),

> The fundamental problems that seem to prevent us from learning from our mistakes and successes in building design could be summarised as follows:
>
> • low image of repair and maintenance
> • lack of academic interest in repair and maintenance
> • lack of organisation among building owners, tenants and property managers
> • lack of strategies to maintain the value of property investments
> • fragmentation of the repair and maintenance industry
> • lack of repetition in design and construction
> • long time cycles between design and repair or improvement
> • poor record-keeping when buildings are constructed
> • lack of standard systems for storing, sorting or accessing construction records
> • poor record-keeping when repairs or improvements are carried out.

Building Pathology as a relatively new discipline provides the intellectual and professional forum for tackling these problems head-on. It can help increase our understanding of the symptoms and causes of building defects and deterioration. It can also act as a useful evidence base for inspecting and surveying buildings. All of this will be needed to reduce the incidence of failure as well as enhance the quality of our building stock and make it more sustainable.

Chapter 2

Principles of building diagnostics

OVERVIEW

This chapter examines the processes involved in diagnosing building defects and failures. It addresses the problem solving, critical thinking and decision making required for the systematic investigation of buildings.

2.1 Introduction to diagnostics

2.1.1 Background

Building diagnostics is the branch of Building Pathology that deals with methodologies and techniques for determining the condition, internal environment and performance of property. It also covers the investigation and commissioning of services within buildings (House and Kelly 2000; Odom and Dubose 2002). It requires the practitioner to adopt a systematic approach and exercise professional judgement when investigating building problems such as defects, poor indoor air quality and lack of performance.

The three main inter-related cognitive branches of building diagnostics are illustrated in Figure 2.1.

There are three main types of building diagnostics, the last of which is the most relevant to the subject matter of this book:

1. **Commissioning diagnostics:** It uses active test signals to force test plant and equipment such as valves, dampers, fans, etc. to a condition where the expected performance of a component or system is satisfactory (House and Kelly 2000).
2. **Monitoring diagnostics:** It involves the passive assessment of components and systems to ensure that they are operating effectively and efficiently (House and Kelly 2000).
3. **Investigating diagnostics:** It involves the systematic examination of a component, element or system that is faulty or not performing adequately to determine its diagnosis and prognosis.

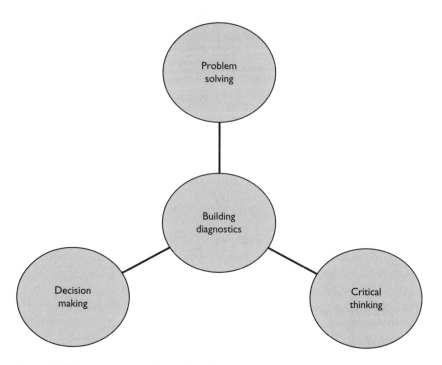

Figure 2.1 The main cognitive branches of building diagnostics.

Diagnosis, the key professional judgement process in building diagnostics, has its roots in medicine. It is derived from the Greek *dia* ("apart", "distinguish", "through") and *gnosis* ("knowledge"). Thus, to put it simply, diagnosis is the art of distinguishing among diseases (or problems) by their signs and symptoms. Signs are those direct clues obtained on site, which are usually visual – such as cracking or distortion of render. Symptoms are those features or indirect clues, which are sometimes non-visual – such as bossing of rendering being symptomatic of detachment – that the client/occupier or maintenance feedback reports about the building.

An initial activity in diagnostics is the need to ask diagnostic questions. In using diagnostic questions one is usually trying to decide whether a test or further investigation is worth doing. The building pathologist may want to know how good a test is at distinguishing elements that have a certain defect or condition from those that haven't got it. A test might be considered better if it is safer, cheaper or less disruptive than another. Such a decision could apply to the use of specialist investigative equipment such as impulse radar or infrared thermography (see Chapter 4).

Broadly speaking, there are five main information needs associated with diagnosis questions in Building Pathology:

1. To judge the severity of a defect.
2. To judge the cause of a defect.
3. To predict the subsequent course and prognosis of a defect.
4. To estimate the likely responsiveness to therapy in the future.
5. To determine the actual response to therapy at the present.

This book is primarily concerned with the first three information needs.

2.1.2 Problem solving

Problem solving is a cognitive activity that attempts to overcome a dilemma or problem by rational means. A great deal of what building pathologists do, like many other professionals, is to solve problems and make decisions. Investigating a defect is clearly a problem-solving process that entails making a decision based on professional judgement as to the likely cause of a defect and the required response.

There are three main categories of problems (Kahney 1993):

1. **Puzzle problems** (e.g. anagrams, crosswords, Su Doku, etc.).
2. **Well-structured problems:** These are also sometimes referred to as transformational problems, which are well defined with a probable solution that possesses a correct, convergent answer (e.g. mathematical calculations).
3. **Ill-structured problems:** These are ill-defined problems because one or more of the problem elements are unknown or not known with any degree of confidence; they can possess divergent solutions, solution paths or no solutions at all.

Building defects typically fall within the ill-structured problem category. This is because they often have no single or clear-cut answer or solution.

Problem solving, therefore, is a process that can involve the strands shown in Figure 2.2.

Solving problems, particularly building defects, is not always straightforward. However, it can be aided by using algorithms or heuristics.

2.1.3 Algorithms and heuristics

An algorithm is a scientific procedure comprising a finite set of well-defined instructions for accomplishing a defined task (i.e. a problem) which will end in a defined state. It is often graphically represented as a flow chart (See Figure 2.3).

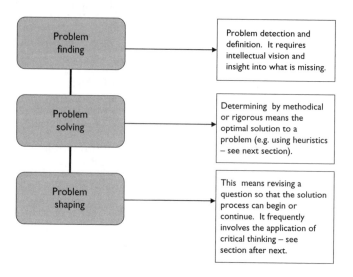

Problem
finding

Problem detection and
definition. It requires
intellectual vision and
insight into what is missing.

Problem
solving

Determining by methodical
or rigorous means the
optimal solution to a
problem (e.g. using heuristics
– see next section).

Problem
shaping

This means revising a
question so that the solution
process can begin or
continue. It frequently
involves the application of
critical thinking – see
section after next.

Figure 2.2 The three strands of problem management.

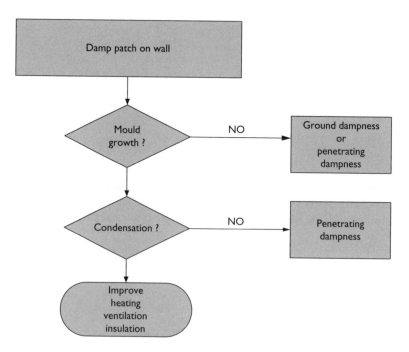

Damp patch on wall

Mould
growth ?

NO

Ground dampness
or
penetrating
dampness

Condensation ?

NO

Penetrating
dampness

Improve
heating
ventilation
insulation

Figure 2.3 Simple flow chart for dampness investigation.

Similarly, a heuristic is a strategy or rule of thumb method of problem solving that is usually designed to provide a speedy and effective solution (Gigerenzer *et al.* 1999). The word is pronounced hyu-RIS-tik and is derived from the Greek *heuriskein*, meaning "to discover" – from which comes, for example, the exclamation term "eureka" (SearchSecurity.com 2005). However, a heuristic cannot always guarantee success – unlike an algorithm (such as a simple flow chart), which usually results in a definite (but not necessarily correct) solution.

The word "heuristic" has uses as an adjective or noun. As an adjective it pertains to the process of gaining knowledge or some desired result by intelligent guesswork rather than by following some pre-established formula.

Heuristic as a problem-solving term seems to have two usages:

1. Describing an approach to learning by trying without necessarily having an organised hypothesis or way of proving that the results proved or disproved the hypothesis. This is basically "seat-of-the-pants" or "trial-by-error" learning. Such an approach is too fickle to be of much use in building pathology or other disciplines that rely on a systematic, rigorous investigation.

2. Pertaining to the use of the general knowledge gained by experience, sometimes expressed as "using a rule-of-thumb". (However, heuristic knowledge can be applied to complex as well as simple everyday problems. Human chess players use a heuristic approach.) As a noun, a heuristic is a specific rule-of-thumb or argument derived from experience. Thus, the application of heuristic knowledge to a problem is sometimes known as heuristics (SearchSecurity.com 2005), which is how it can be used in Building Pathology.

A simple heuristic (i.e. problem-solving tool or strategy) is described in Table 2.1. It is based on the statement: "How to make a **DENT** in a problem: Define, Explore, Narrow, Test" (source: http://www.saltspring.com/capewest/pbl.htm - Accessed 12/02/06).

The aim of using a heuristic for diagnosing defects is to help the novice surveyor gain the minimum level of competency quicker than would otherwise be the case. Initially, it can be quite useful to help the novice surveyor focus on or prioritise the likely suspects or culprits.

However, algorithms and heuristics must be used with caution. They are only tools to aid in problem solving. If accepted uncritically when judging risk and uncertainty they can lead to misperceptions of risk and poor decisions (Kahneman *et al.* 1982). That is why critical thinking is important in building diagnostics and other investigative disciplines.

Table 2.1 The DENT problem-solving process

Key stages	Actions
Define	• Carefully define the problem. • Detect/discover any anomalies. • Cue acquisition – of possible defect source or agencies involved.
Explore	• Hypothesis generation. • Devise a list of knowledge of the problem. What do we know? What do we don't know? Is this defect analogous to any past problem? • What core building concepts may apply to this problem?
Narrow	• Cue interpretation of evidence. • After developing a list of hypotheses, sort them, weed them and rank them. List the type of data required to test each hypothesis. Give priority to the simplest, least costly tests. It is easier to get information on the condition of a building than it is to get from sophisticated laboratory tests on materials.
Test	• Hypothesis evaluation. • When you encounter data that confirm one of your hypotheses draft a technical report giving an explanation of your solution and justify it using the available evidence. • If all your possible solutions are eliminated, begin the cycle again: define, explore, narrow, test.

2.1.4 Critical thinking

Effective and efficient problem solving and professional judgement requires critical thinking. It's the cognitive link between diagnostics and best practice (see Figures 1.1 and 2.1). Critical thinking, in other words, is another cognitive process, which involves assessing information, especially claims or propositions that people assert as true (Cottrell 2005). This means taking nothing for granted. A critical thinker, therefore, identifies and challenges assumptions, considers what is important in a situation, imagines and explores alternatives, and applies reason and logic to make informed decisions (Potter and Perry 2001). In other words, critical thinking means seeking *reliable* knowledge. It therefore forms a process of *reflecting* upon the meaning of statements, examining the offered evidence and reasoning, and forming judgements about the facts. (See website at www.criticalthinking.com for further guidance on this important part of diagnostics.)

Critical thinkers can gather such information from: observation, experience, reasoning and communication. Good critical thinkers face problems without jumping to a speedy, single solution. Instead they focus on the options for what to believe and do (Potter and Perry 2001).

Critical thinking has its basis in intellectual values that go beyond subject-matter divisions and which include clarity, accuracy, precision, evidence, thoroughness and fairness. Other attitudes relevant to critical thinking are confidence, thinking independently, responsibility and authority, risk taking, discipline, perseverance, creativity, curiosity, integrity and humility (Potter and Perry 2001). The key skills in this regard are summarised in Table 2.2.

Table 2.2 Critical thinking skills proposed by the American Philosophical Association (based on Potter and Perry 2001)

Skill	Description	Building pathology applications
Interpretation	• Categorisation • Decoding sentences • Clarifying meanings	Be systematic in data collection. Look for patterns to categorise data. Clarify any data you are uncertain about.
Analysis	• Examining ideas • Identifying arguments • Analysing arguments	Be open-minded as you look for information about a defect/building. Do not make careless assumptions. Do the data reveal what you believe is true, or are there other options?
Evaluation	• Assessing claims • Assessing arguments	Look at all situations from an objective view. Use criteria (e.g. expected outcomes) to determine the results of any actions or interactions. Reflect on your own behaviour.
Inference	• Examining evidence • Speculating or conjecturing alternatives • Making conclusions	Look at the meaning and significance of findings. Are there relationships between findings? Do the data about the defect/building help you in seeing that a problem might exist?
Explanation	• Stating results • Justifying procedures • Presenting arguments	Support your findings and conclusions. Use knowledge to select strategies you use in the care of buildings.
Self-regulation	• Self-examination • Self-correction	Reflect on your own experiences. Identify in what way you can improve your own performance. What will make you feel that you have been successful?

The key components of critical thinking are specific knowledge base, experience, competencies, attitudes and standards. These underpin the following three levels of critical thinking (Potter and Perry 2001):

- Level 1: Basic – application of principles and adherence to standards of critical thinking;
- Level 2: Complex – independent and evolving critical thinking; and
- Level 3: Commitment – proactive and accountable critical thinking.

Given the above, it is clear that effective problem solving requires critical thinking and an orderly approach. Novice or inexperienced practitioners do not magically acquire problem-solving skills by simply having tutors or experienced practitioners throwing problems at them. They require a methodical approach that develops their critical thinking as well as decision-making skills and the use of heuristics. This edition aims to provide guidance in this area.

2.1.5 Scientific method applied to defects diagnosis

The basic framework for any systematic enquiry is the scientific method. The main steps in this process are illustrated in Figure 2.4.

Building diagnostics or defects investigation involves decision making (see next section) as well as problem solving. This process develops from the scientific method shown in Figure 2.4, and is summarised in Figure 2.5. It follows the well-established procedure adopted by medical clinicians (Barrows and Pickell 1991). The generation of a hypothesis is a critical step in the process. As indicated earlier, a hypothesis is an assertion or proposition that one can test (e.g. My initial judgement is that condensation is the cause of the dampness problem). The building pathologist then investigates this hypothesis to determine one of three possible outcomes:

1. The hypothesis is true (e.g. condensation is indeed the cause of the dampness problem).
2. The hypothesis is false (e.g. penetrating dampness not condensation was the cause of the moisture problem).
3. The hypothesis cannot be confirmed or confuted – usually owing to lack of evidence.

2.2 Decision making in defects diagnosis

2.2.1 Basics of decision making

The cognitive process of selecting a course of action from among two or more alternatives is known as decision making. It's a process that should

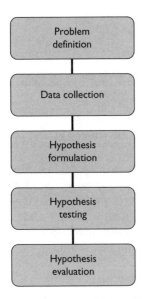

Figure 2.4 Steps of the scientific method (based on Potter and Perry 2001).

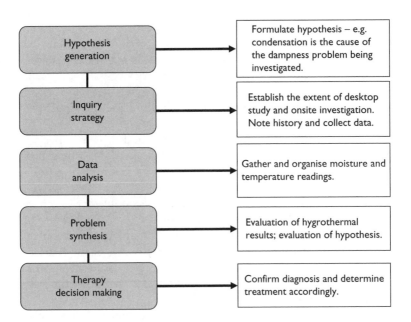

Figure 2.5 Problem-solving, decision-making process applied to defects diagnosis (based on medical model by Barrows and Pickell 1991).

allow for feedback at every stage, and is similar to problem solving in that it involves the following:

- First, identify and define the problem.
- Second, develop alternative hypotheses (i.e. tentative statements that can be tested).
- Third, evaluate alternative hypotheses.
- Fourth, make and (if satisfied) implement the decision.
- Fifth, evaluate and monitor results (and provide feedback).

In most decision-making cases a final choice is produced. The result is usually an action or opinion. The process begins when we need to do something but we do not know what. Therefore, decision making is a reasoning process which can be rational or irrational, and can be based on explicit assumptions/knowledge or tacit assumptions/knowledge.

Explicit knowledge can be derived from the following sources:

- *Best practice*: Authoritative sources such as the BRE
- *Business function*: Expertise of the consultancy firm concerned
- *Clients/customers*: Printed briefs, written instructions
- *Corporate knowledge*: Specialist expertise available
- *Product information*: Manufacturers' catalogues/reports, certifying bodies such as the British Board of Agrément.

Tacit or implicit knowledge, on the other hand, is not always as accessible or conspicuous. It can comprise the following:

- *Departmental/organisational databases*: libraries, previous projects' reports.
- *Existing infrastructure*: Buildings and sites, and their histories.
- *People*: Past experience, familiarity with a specific building problem, feedback on the building's history.

In everyday life we make decisions, common examples of which include shopping, selecting what to eat, and deciding whom or what to vote for in an election or referendum. Defects diagnosis involves decision making in that it requires the practitioner to determine which of the two or more causes/treatments would be respectively the most accurate/appropriate. Decision making involves choosing an action after weighing the risks and benefits of the alternatives.

Decision making, however, is essentially a psychological construct (Plous 1993). This means that although we can never "see" a decision, we can infer from observable behaviour that a decision has been made. Therefore, we conclude that a psychological event that we call "decision making" has occurred. It is a construction that imputes commitment to action. That is, based on observable actions, we assume that people have made a commitment to effect the action.

Once a building pathologist receives information about a defect or property with a problem in a particular situation, diagnostic reasoning begins. It's a process of determining for a client a building's state of repair or performance. The following are the factors influencing decision making in this regard:

- Potential benefits of diagnosis
- Potential risks of diagnosis
- Uncertainty in diagnosis
- Costs of investigation
- Utility of diagnosis.

Diagnostic reasoning enables the practitioner to assign meanings to anomalies, physical signs and reported building problems. This process entails a series of professional judgements made during and after data collection of the defect investigation, which results in an informal judgement or formal diagnosis. It also assists in establishing inferences or judgements about a defect's prognosis.

Structured rational decision making is an important part of all science/evidence-based professions, where specialists apply their knowledge in a given area to make informed decisions. For example, medical decision making often involves making a diagnosis and selecting an appropriate treatment. Like the health-care professions, legal practitioners can also use decision analysis; the latter can use it to counsel clients on their litigation risks and help them negotiate optimal settlements (www.treeage.com/learnMore/index.html).

Decision making in defects diagnosis can be categorised into two levels. These are illustrated in Figures 2.6 and 2.7.

2.2.2 Decision analysis

Decision analysis is the application of explicit, quantitative methods for analysing decisions under conditions of uncertainty. It allows practitioners to compare the expected consequences of pursuing different strategies with the information they need to make informed decisions now. The process of decision analysis makes fully explicit all of the elements of the decision, so that they are open for debate and modification. While a decision analysis will not solve a practitioner's problems, it can help them explore their decision.

Decision analysis is a more consistent, structured and rigorous decision-making strategy. In literature analyses, for example, it involves two key criteria: sensitivity and specificity. The first is useful in getting the right result, whilst the latter is useful in avoiding getting the wrong

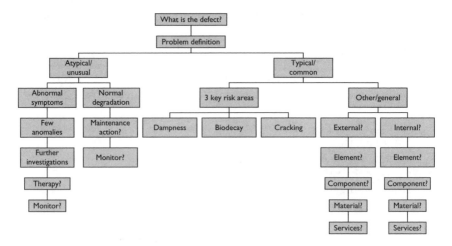

Figure 2.6 First level of decision making in defects diagnosis.

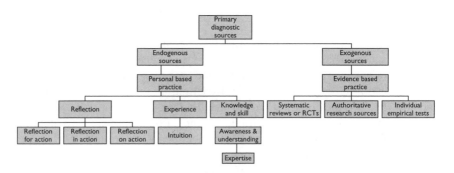

Figure 2.7 Second level of decision making in defects diagnosis.

result when searching for information. According to an NHS source (http://www.elib.scot.nhs.uk/portal/Upload/Librarian/Unit3dx.doc),

> In searching literature, these two terms have an analogous meaning to the way they are employed in diagnosis. Specificity refers to the proportion of data or documents that you retrieve which are relevant; while sensitivity refers to the proportion of all the documents which are relevant that your search manages to retrieve. (These terms are also sometimes called **precision** and **recall** respectively).
> Put another way:
>
> • Sensitivity is the likelihood of retrieving relevant items;
> • Specificity is the likelihood of excluding irrelevant items.

If you're doing a search and it yields an unmanageably large number of hits, you probably need to increase the specificity of your search; conversely, if you get too small a number of hits, you probably need to increase sensitivity.

Increase specificity by:

- narrowing your question!
- using more specific terms in Free Text search
- using more specific Thesaurus/Subject headings
- adding in terms (using AND) to represent other aspects of the question
- limiting to language of article, to material or component subjects, to publication types (e.g. randomised controlled trials, reviews, etc.), to country or year of publication.

Increase sensitivity by:

- broadening your question!
- finding more search terms from relevant records
- trying different combinations of terms
- adding in and combining terms of related meaning using OR
- searching further back in time

When there are no negative options, and the choices are clear, consequences are minor, and the decision rests with a single decision maker, it's relatively easy to make a sound decision. However, in Building Pathology and many other disciplines, especially medicine, decisions often entail relatively high stakes. This is because of influences such as uncertainty, competing priorities, multiple decision makers and limited information. Frequently, situations can entail high-impact decisions involving human safety, significant investment, high complexity and elements of uncertainty. In such cases, a sound decision-making process will help in giving practitioners a better chance of achieving a positive outcome – i.e. in the context of a building defect investigation, a correct diagnosis (see http://www.treeage.com/learnMore/index.html).

One of the method's virtues is that it allows decision makers to quantify the uncertainties involved in a decision by expressing them in terms of probabilities. A diagnosis with a 0.95 (i.e. 95%) probability of being correct is at face value cogent. This distinction helps ensure the inclusion of some factors that decision makers might otherwise be tempted to dismiss.

Decision analysis makes use of software tools (e.g. *Criterion Decision Plus* or *TreeAge Pro*) to help users build models that represent specific decision problems. It then relies on statistical analysis to either determine the best course of action, or to discover what information is required to make a good decision.

Mapping alternatives on a fault tree model helps illustrate all of the factors affecting a decision and each of the possible outcomes for decision makers. In some situations, delaying a decision can be more costly than embarking on a less than perfect path.

The difficulty with complexity for decision makers such as building pathologists is that it can often lead them to concentrate on the wrong problems. However, by incorporating a process called sensitivity analysis, decision analysis can help determine which factors are really critical to a decision's outcome. This technique is used in the medical profession and other disciplines involving high stakes. Such knowledge keeps decision makers from focusing on distinctions that make little difference, helping them avoid "analysis paralysis" and move on to a final decision.

A good decision, of course, is never guaranteed when using decision analysis. As with other heuristics, it's dependent on the quality of the information used – the "garbage in, garbage out" rule applies. Still, "when it is well executed, incorporating probabilities based on accepted data and expert opinion, decision analysis generates highly credible results" (http://www.treeage.com/learnMore/index.html).

The use of sensitivity analysis can be illustrated in the following scenario involving a dampness problem in a stock of dwellings. It is based on a medical case study presented by Dr MW Dawes (Source: www.cebm.net/downloads/april_01/Decision_Analysis.ppt).

Scenario

A local authority owns and maintains 250 traditional two-storey semi-detached dwellings. A recent local house condition survey of the stock found that the majority were affected by dampness. Investigators narrowed the culprits down to two possible causes: condensation and penetrating dampness. A decision analysis exercise was used to help determine the most likely source and required treatment for this problem.

Process

(a) Problem

- What would happen if you opted for one path in preference to another?
- On what basis would you assess outcome?

(b) Start solution

- List all the options and display them.
- Commonly accepted format is a decision tree diagram. This is a graphical tool to aid decision analysis/making. A decision tree for

the diagnosis of defects can be found in Carillion Services (2001). (See Figures 2.8–2.11 for examples of simple decision trees that can be used to illustrate this exercise.)

(c) Decision data

- Condensation/Penetrating dampness
- Treat/Don't treat
- Result: 0.9 better than 0.1.

(d) Add reality to probability scores

- Cost of tests
- Cost of treatment
- Days lost from work
- Cost of re-attending investigator
- Multiply the probabilities by the costs.

(e) Determine costs data

- Rollback costs are = probability × estimated costs
- Results: More dwellings get better (90% vs 20%) and it is cheaper (2.70 vs 7.60).

(f) Main decision

- Condensation is the more likely cause of the dampness problem.
- Therapy to tackle condensation is an effective and economical solution.

2.2.3 Other decision-making techniques

It is beyond the scope of this book to examine the many other decision-making techniques available. Moreover, not all of them would be suitable for helping to diagnose defects. However, the website www.mindtools.com contains useful information on the following:

- Pareto analysis (see also Manktelow 2003)
- Six thinking hats (see also deBono 2003)
- Positive Minus Interesting (PMI)
- Grid analysis
- Paired comparison analysis
- Force field analysis
- Cost/benefit analysis.

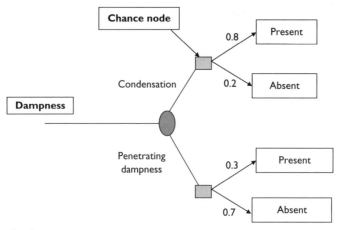

Result:
0.8 is better than 0.3

Figure 2.8 Simple decision tree to determine the more likely cause of a dampness problem.

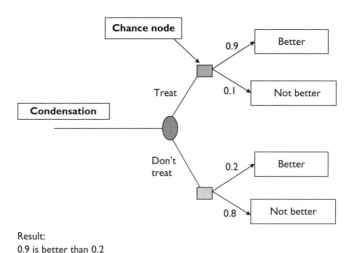

Result:
0.9 is better than 0.2

Figure 2.9 Simple decision tree evaluating treat versus do not treat a dampness problem.

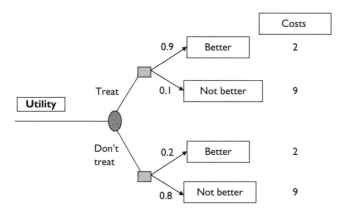

Figure 2.10 Simple decision tree showing costs of treating versus not treating a dampness problem.

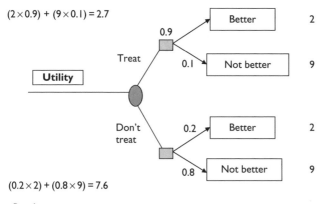

$(2 \times 0.9) + (9 \times 0.1) = 2.7$

$(0.2 \times 2) + (0.8 \times 9) = 7.6$

Result:
- More dwellings get better (90% vs 20%)
- It is cheaper (2.70 vs 7.60)

Figure 2.11 Simple decision tree evaluating costs of treating against not treating a dampness problem.

2.2.4 Post-diagnosis decisions

After a final diagnosis is made (which is, hopefully, correct) about a defect, the investigator is faced with a number of options. These can be summarised as follows:

- Do nothing
  Unless the defect is very minor, doesn't affect building safety or performance, or is highly unlikely to get worse, this is usually not a sensible option.

- Watchful waiting
  This is a term used in healthcare, and means monitor the patient/situation (in a construction context via regular maintenance inspections) to see if the disease/defect gets any worse. This ought to be the basic option if none of the following are feasible.

- Treat the symptoms
  In some cases, neither watchful waiting nor a full cure is appropriate. In order to relieve the effects of a building problem (e.g. mould growth on walls) it may at least be worthwhile to treat the symptoms. However, as this doesn't resolve the root cause, the problem is unlikely to go away. It's more likely to get or come back worse than before.

- Cosmetic treatment and control
  This option should be seen as a last resort before contemplating doing nothing. In some cases where a full cure is not possible, hiding the effects of the defect might be deemed desirable. For example, in basements of some old buildings it may be impossible to eradicate water ingress completely. A panelled wall incorporating a drained tanking system covering the damp walls can offer a suitable "remedy" in such circumstances.

- Further investigations
  In some cases, a full diagnosis of a defect, based on the previous investigation, is not possible. This could be for a variety of reasons – such as insufficient time, lack of relevant data and poor quality of evidence. In these circumstances it would be prudent for the investigator to recommend appropriate further investigations on the defect concerned if it's serious enough.

- Full cure
  This is clearly the ideal option, which minimises the risk of the problem reoccurring. But risk can never be entirely eliminated from Building Pathology, and so it is to risk we now turn our attention to.

2.3 Risk in building diagnostics

2.3.1 Significance of risk

Risk generally has a negative connotation in that it usually involves some form of financial loss. It has also been defined as the combination of the possibility of an event and its consequence (BSI 2002a). Alternatively, risk can be seen as the chance that an actual outcome will deviate from that forecast or intended.

Closely related to risk is uncertainty. The main difference between them is that the former is something that is considered to be reasonably objective in nature and thus quantifiable, whilst the latter is more subjective but generally unquantifiable. Risk always involves an element of uncertainty, but uncertainty does not always involve risk (Douglas 2006).

The differences between risk and uncertainty are illustrated in Figures 2.12 and 2.13.

Figure 2.12 can be contrasted with the risk – uncertainty spectrum shown in Figure 2.13.

The degree of "uniqueness" of a building can influence the degree of associate risk and uncertainty associated with its defects. This is determined by the homogeneity and heterogeneity of the building's materials, components and elements.

Uncertainty	Risk	Certainty
Spectrum of risk ⬅━━━━━ ━━━━➡		
Unknown Unknowns	Known Unknowns	Known Knowns
No information	Partial information	Complete information

Figure 2.12 The basic risk spectrum (based on Bowles and Kelly 2005).

Risk	Uncertainty
Quantifiable	Non-quantifiable
Statistical assessment	Subjective probability
'Hard' data	Informed opinion

Figure 2.13 The risk – uncertainty spectrum (based on Douglas 2006 and Raftery 1994).

There is less uncertainty and thus potentially lower risk associated with homogeneous characteristics of a building. This occurs where the elements of a building are the same or similar (e.g. timber: roof, walls and floors).

Heterogeneity, on the other hand, is a more distinguishing characteristic of all buildings. It is this which creates the conditions of greater uncertainty about the outcome of events or situations on sites. In consequence, there is a higher degree of risk with construction work. According to Bowles and Kelly (2005), it results in a range of project differences such as:

- Building and site conditions:
  No two buildings have the same exposure conditions, performance levels, wear and tear or intensity of use.

- Element specifications:
  The actual specification and design of new (e.g. replacement) and existing elements is wide and varied.

- Personnel and style of management:
  Personnel compositions and levels are never the same, and the site/project management style adopted can vary from job to job.

  Another term related to risk is "hazard". A hazard is anything that can cause harm – such as chemicals, electricity, working from ladders, etc. It can result in the risk of physical damage, injury or death. Risk is the chance, high or low, that somebody will be harmed by the hazard (HSE 2003a).

2.3.2 Risks of misdiagnosis

Misunderstanding the causes of a defect can easily result in misdiagnosis, which in turn often leads to mistreatment or unnecessary remedial works. The following are typical examples involving mistreatments relating to the three key defect risk areas:

1. Retro-fit DPC to combat alleged rising dampness instead of treatment for condensation or bridging dampness.
2. Wet-rot repairs instead of dry-rot treatment.
3. Underpinning for suspected subsidence instead of repairs to cracks caused by hygrothermal factors such as thermal movement.

The first main risk of misdiagnosis is that it can easily induce no action if the problem is not considered severe enough. It is more likely to result in mistreatment or unnecessary repairs if some active response is accepted as being necessary. This can easily initiate litigation in the event of the inevitable reoccurrence or worsening of the defect. Litigation is of course an expensive, stressful and time-consuming process that might mean damages being awarded against the defendants, which in turn could undermine their credibility or reputation.

2.4   Summary

Diagnostic reasoning begins as soon as the building pathologist receives information about a defect. It's a decision-making process that involves problem solving, critical thinking and decision analysis. Like any such process, it cannot guarantee a successful outcome. However, adopting the procedures indicated in this chapter and the next can help minimise the risk of misdiagnosis.

Chapter 3

Basic investigative methodology

OVERVIEW

This chapter examines the processes involved in investigating building defects and failures. It shows how to apply diagnostic skills required for the rigorous investigation of building problems outlined in the previous chapter.

3.1 Background

Identifying the precise causal factor(s) of building defects is frequently an onerous task for any investigator (Addleson 1992). Constraints of access, time and resources often make the investigator's job difficult and demanding. When problems occur in buildings it is crucial that their root cause is properly identified. If the initial diagnosis is inadequate or mistaken, this may result in

- the significance of the defect being underestimated, leading to it being ignored, unattended or marginalised;
- the significance of the defect being overestimated, leading to excessive or unnecessary remedial work;
- the wrong or inappropriate repair leading to the defect not being effectively cured/rectified, which could even aggravate the problem.

Clearly, considerable care is required if scarce resources in terms of labour-hours, money and materials are not to be wasted on inappropriate or unnecessary repair work triggered by the wrong diagnosis. The proper identification of the cause(s) is also essential where the liability for the cost of remedial work is to be reasonably ascertained.

Every type of building contains defects to some degree. The consequences of such defects may be minor, but others are more important and may affect the building's appearance and usage of the property. However, the cost of rectifying or (worse still) not rectifying building defects could be extensive. In more serious instances, they may pose a hazard to health and safety for those in and around the property.

In many cases, however, establishing the precise causal factor(s) of a building defect is neither simple nor straightforward. There are several reasons for this. It is not easy to dismantle fully the part of the building where a defect has appeared to expose fully the extent and nature of the problem. This is complicated by the all too often incomplete nature of information about the affected element or its history. Documents such as the original specification or drawings may be missing or only partially complete. A full breakdown description of the make-up of the material(s) originally used in a building is often unavailable. In addition, for obvious commercial reasons suppliers are often unwilling to divulge the precise composition and manufacture of their materials. This often makes it tricky to achieve the correct diagnosis without a thorough investigation.

Thus "the process of diagnosing faults in buildings is often made difficult in a specific situation because of the presence of relatively few anomalies (or symptoms) but with a wide range of possible causes and defects" (Addleson 1992). Moreover, the precise trigger mechanism (i.e. agency) of a building fault is not always apparent even though the cause (e.g. dampness) may be obvious. The same symptoms can be caused by different agencies. A variety of agencies working in combination can trigger a defect. For example, rainwater penetration can lead to dry rot, which is the most destructive form of timber decay in buildings. Dampness is of course the "cause" of the problem in this case, but the source of the moisture may not be so obvious.

Table 3.1 Some common diagnostic techniques

Technique	Example
Case histories	Examining reports, correspondence and other documentation concerning the defect in question.
Pattern recognition	Identifying the signs and symptoms of the defect to determine how well they fit with the cause of other similar defects.
Hypothetico-deductive reasoning	Prescribing a possible cause/s and rigorously testing the evidence by deduction (general to particular) to see if it supports that tentative judgement. This method helps to contain what would otherwise be an infinite search (Barrows and Pickell 1991)
Inductive reasoning	Reasoning from the particular to the general – the opposite of deductive reasoning. Gathering data, analysing them, synthesising them to see if they point towards any specific cause. This is similar to pattern recognition.
Fault/Decision tree analysis	Using decision analysis diagram to determine the most likely defect or one with the highest probability (see Figures 2.5–2.8).
Other heuristic tools	Using the key defects risk model supported by an appropriate checklist as described in Appendix B.

In any therapy to eradicate fungal attack, identifying and eliminating the source of moisture is a primary control measure.

Similarly, a single cause can result in several symptoms. For example, penetrating dampness through a blocked cavity wall can result in corrosion of the wall ties (leading to horizontal expansion of the mortar joints in the outer leaf) and moisture staining or soiling on the inside face of the inner leaf of brickwork.

When a problem occurs in a building it is crucial that the root cause is identified quickly. If the initial diagnosis is inadequate or erroneous, this may lead to the defect not being effectively cured. It is also likely to result in a reoccurrence or aggravation of the problem.

Building pathologists can use one or a combination of the techniques listed in Table 3.1 when investigating defects.

3.2 General investigative procedure

3.2.1 Standard procedure

The essential frameworks for investigating defects are shown in Figure 3.1. They are similar to that for undertaking a building survey (Noy and Douglas 2005). The process consists of the following key stages:

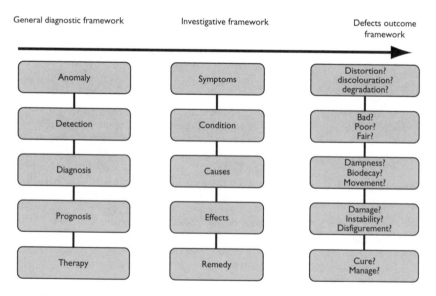

Figure 3.1 Essential frameworks of the process for investigating building defects.

3.2.2 Problem definition

(a) Establish the scope of the investigation.

- Limited to a specific material?
- Restricted to a component or element.
- Preliminary or full investigation required?

(b) Establish the nature and extent of the defect.

- Leak or other moisture problem?
- Degradation or deterioration of material or component?
- Isolated or widespread problem?

(c) Identify the condition and any anomalies (adverse symptoms) relating to the affected element.

- Building dilapidated?
- Building otherwise in excellent condition?
- Changes in appearance?

3.2.3 Desktop phase

- Compile background information on the building and/or the defect under investigation.
- Consult documents such as previous reports, correspondence relating to the problem.
- Check as-built drawings if available.
- Check legal documents such as title deeds, building control approval, etc.
- Determine extent of equipment and tools needed for the proposed investigation.

3.2.4 On-site phase

- *Reconnaissance* – ascertain physical context of property to check for common problems and site risk factors.
- *External preview* – inspect condition of building/element being investigated.
- *Internal inspection* – carefully investigate the locus of the anomaly to trace the internal extent of the defect.
- *External review* – check condition of suspect area/element to confirm internal inspection; preliminary diagnosis.

3.2.5 Reporting phase

- Confirm original instructions.
- Describe methodology used in the investigation.

- Data collection, presentation, analysis and discussion.
- Conclusions – confirm/indicate likely diagnosis; indicate prognosis – outlining the consequences of the defect, especially if it is not remedied.
- Recommendations – suggest immediate remedial action, including, if necessary, advice on procurement of repair works; indicate if further investigations are needed.

3.2.6 Monitoring phase

- Establish frequency and form of monitoring of area affected by the defect.
- Make regular checks on chosen therapy to determine its efficacy.

3.3 Diagnostic process

3.3.1 Essential steps

Several methodical steps in the decision-making process should be undertaken to achieve a successful (i.e. accurate and prompt) diagnosis. This procedure is illustrated below in considering a dampness problem.

- *Problem*: Damp patch on the inside face of an external solid wall in the ground floor bedroom of a mid-terraced, two-storey dwelling.
- *Procedure*: See Table 3.2.

3.3.2 Analysis and interpretation of data

Key pieces of evidence in any investigation are data (facts, figures and information). These are initially collected from various relevant sources and validated (Potter and Perry 2001). Data analysis involves recognising patterns or trends, comparing them with normal building performance standards, and arriving at a reasoned judgement about the building's response to a defect. When searching for a pattern or trend, the building pathologist examines the data in the database as shown in Table 3.3.

A set of signs or symptoms that are grouped together in a logical order is called a cluster (Potter and Perry 2001). It is this pattern that emerges. On their own these clues don't tell the building pathologist enough to make a diagnostic judgement – just a hypothesis. When these signs are placed or clustered as a group, however, the building pathologist can see a relationship between and among these assessment findings.

Defining characteristics of a defect are contained within clusters and patterns. The performance or diagnostic criteria or assessment findings

Table 3.2 Defects diagnosis procedure

Stage	Example
1. Propose a hypothesis for the cause (i.e. make preliminary diagnosis) and test it. Establish if possible whether or not the defect is a typical or atypical one.	This is the first level of decision-making in defects diagnosis, and it's illustrated in Figure 2.3.
2. Consider the presence of relevant evidence and disregard irrelevant evidence.	• Relevant evidence: damp wall below 1m. • Irrlevant evidence: dry wall above 2 m.
3. Note the absence of possibly relevant evidence.	• No tide mark or efflorescence. • No fungal attack to adjoining timbers.
4. Consider the range of diagnostic options to confirm/confute hypothesis or suggest an alternative.	Further investigations: • visual inspection only. • visual inspection and on-site test. • visual inspection and off-site test.
5. Test the evidence and/or absence of evidence against the options.	• This is the second level of decision making in defects diagnosis, and it's illustrated in Figure 2.4. • Confirm/confute hypothesis.
6. Make a considered choice as to the likely cause (i.e. a firm diagnosis).	Consider the implications of any discrepancies between the evidence and absence of evidence and your considered choice.
7. Identify the level of confidence in your diagnosis.	95% (0.9) confidence level = very high.

that support (i.e. validate) the presence of a diagnostic category are the defining characteristics. Performance or diagnostic criteria are objective or subjective signs and symptoms, clusters of signs and symptoms, or risk factors. The presence of multiple defining characteristics as a result of data assessment would tend to support the diagnosis. In contrast, the absence of these features indicates that the proposed diagnosis should be rejected. When all characteristics are assessed, non-relevant ones are eliminated and relevant ones are confirmed, precision is achieved (Potter and Perry 2001).

Data in defects diagnosis can come in several forms. Table 3.3, for example, illustrates the types of data required for a dampness investigation.

Table 3.3 Types of dampness data

Type of data	Example
Moisture content	• Surface readings? • Core (or subsurface) readings? • Air humidity?
Type of moisture in masonry	• Capillary? • Free? • Hygroscopic?
Temperature	• Ambient air temperature? • Surface temperature? • Dew point temperature?
Organic growths	• Algae/lichen/moss? • Mould (species)? • Fungi (type)?
Tide marks	• Regular pattern? • Irregular pattern?
Volume of water	• High: leaking services; flooding. • Medium: severe condensation; rainwater penetration. • Low: condensation; bridging/rising dampness.
Salts	• Chlorides? • Nitrates? • Others?
Damage	• Corrosions of metals? • Deteriorated finishing? • Efflorescence/mould on wall/ceiling surfaces? • Fungal attack to timbers? • Indentation/breakage of building elements?

3.3.3 Types of diagnoses

The building pathologist is ready to formulate defects diagnoses once patterns and clusters of data are sorted and client needs are identified. The following is a list of the main types of defects diagnoses (based on Potter and Perry 2001):

- *Actual defect diagnosis*: A diagnosis that is professionally validated by the presence of major defining characteristics (e.g. surface condensation confirmed as a result of extensive mould growth and relative humidity levels above 70%).
- *Risk defect diagnosis*: This describes human responses to building conditions that may develop in vulnerable components, elements or buildings.
- *Tentative defects diagnosis*: This describes a suspected problem for which current and available data are insufficient to validate the

problem – for example, in a complex dampness investigation where there are several potential sources of moisture.

- *Syndrome diagnosis*: A diagnostic label given to a cluster of building problems that frequently go together and present a picture of some anomaly – such as sick-building syndrome.
- *Condition or performance diagnosis*: Judgement about the state of repair or overall performance of a building.

3.3.4 Detecting defects

Diagnosing defects is rarely a straightforward activity with few risks. As indicated earlier, uncertainty, lack of knowledge and skill, and insufficient time can all inhibit if not prevent an investigator from achieving a correct diagnosis.

Detection of a defect is usually undertaken by identifying certain indicators using most if not all of the five human senses as outlined in Table 3.4. However, these have their limitations in terms of accuracy, sensitivity and consistency. For these reasons building pathologists often have to rely on analytical tools and techniques (see Chapter 4).

Table 3.4 Human senses and their use in detecting defects

Faculty	Example
Visual	Look for changes in appearance: • staining • discolouration • indentation • distortion or irregular pattern.
Olfactory	Try to smell for unusual or foul odours: • mushroomy smell is indicative of dry rot; • obnoxious smell may be indicative of either a defective foul drain or a dead animal rotting in a hidden space.
Aural	• Detect bossed render by tapping it – a hollow or dull sound is indicative of bossing; a clear, sharp noise is indicative of sound substrate. • Hear problems such as water hammer using a listening stick (Burkenshaw and Parrett 2003). • Listen for dull sound when tapping timber with a hammer to ascertain if it is affected by fungal or insect attacks.
Tactile	• Feel surface for unevenness, roughness or loose material (e.g. powdering and flaking of paint). • Feel material for friability (i.e. the crumbly state of mortar or wood is indicative of a problem).
Taste	Bitter taste in drinking water may be indicative of contamination. However, the safe use of this faculty is limited because of the risk of contamination or poisoning.

3.3.5 Cause determination

As indicated earlier, establishing the root cause of a defect or failure is fundamental to arriving at a correct solution (Figure 3.2). Tracing this is often easier said than done. Most buildings, for example, cannot be easily dismantled to facilitate inspection of their hidden parts and then reassembled. Some invasive work, which is often expensive and disruptive, is often required when investigating a defect. For example, removing part of a floor or ceiling to expose joist ends to check for fungal attack is a messy and damaging operation. This of course can be especially problematic if the building is Listed or in a Conservation Area. However, such invasive work may be prudent if there is any doubt about the presence of defects such as dry rot.

According to the CIB (1993) three types of cause descriptions are commonly recognised as indicated in Table 3.5 below.

The priorities given to each of these cause descriptions will depend on the exigencies of the job and the requirements of the client. Technique determinations are likely to come first, however, so that a solution is derived as quickly as possible. Liability-orientated determinations may follow if litigation is an issue. In the long run, system-orientated determinations will be important as part of any "input" for quality assurance in the building process (CIB 1993).

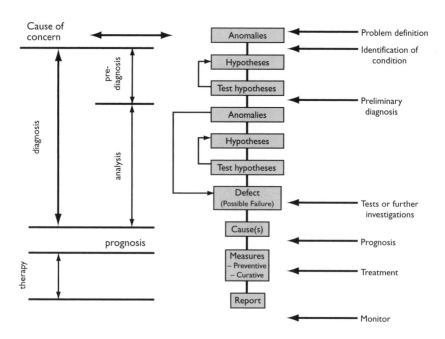

Figure 3.2 Defect analysis procedure (adapted from CIB 1993).

Table 3.5 Cause determinations

Determination	Key question	Consequence
Technique determinations	What caused the defect? • Agency/mechanism • Trigger?	Therapy? • Cure? If possible. • Manage? If not possible to cure.
Liability-orientated determinations	Who caused the defect? • Designer/specifier? • Constructor? • Occupier? • Other?	Culpability? • Dispute? • Litigation? • Damages?
System-orientated determinations	How did the defect originate? • Source – in walling/roofing, etc.? • Bad workmanship? • Faulty design? • Inferior material?	Quality? • Upgrade specification? • Improve supervision?

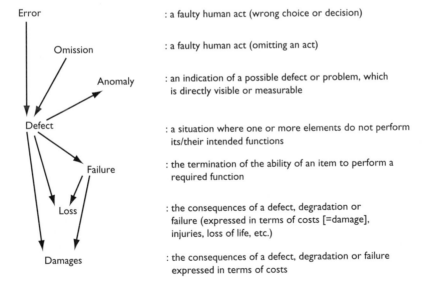

Figure 3.3 Chain of events (CIB 1993).

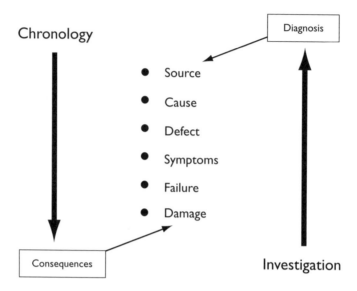

Figure 3.4 Sequence of investigation.

The chronology of a building problem starts with a source or error, which then gives rise to a cause, leading to a defect, which is manifested as an anomaly, then failure, loss and damage (Figure 3.3). In some respects, the investigative process works in reverse to this sequence (Figure 3.4). It is the gathering and sifting of the evidence (to establish extent of data validation and data clustering) to arrive at a diagnosis that leads to the correct detection of the cause and source of the problem.

3.4   Models of analysing defects

3.4.1   Simple models

Models are representations that serve to explain concepts (Potter and Perry 2001). There are a variety of models that can be used as the basis of procedures and methods for investigating building defects. One is the key defect risk areas model outlined in Section 3.4.6.

This model is incorporated in a brief checklist covering the main elements and components of a building (see Appendix 2). For more complex cases, diagnostic charts, matrices, fault trees and even expert systems can be used as diagnostic aids (CIB 1993).

All the indications are therefore that a careful, systematic or forensic method is a more effective way of investigating, assessing and diagnosing building problems than adopting a random or haphazard approach. Using the latter might result in a correct diagnosis but is likely to prove less consistent and more unreliable than the former.

According to Smith's (1987) experience of investigating defects, it has become clear that the diagnostic path has two parts:

1. *First question path*: A sequence of questions that lead to a firm diagnosis as to the type of defect (e.g. type of dampness).
2. *Second question path*: A sequence of questions that lead to the identification of the defect that caused the problem (e.g. source of dampness).

A general framework for investigating defects requires, therefore, the building pathologist to ask particular questions. The following basic diagnostic questions were propounded by Smith (1987):

- **What are we diagnosing?**

Often there may be more than one factor responsible for the existence of a problem, and the factors may change when the same problem arises in different circumstances. But the conclusions drawn must be nonetheless specific.

- **How is the problem to be classified?**

It seems certain that the most useful classification is in terms of what causes the problem, since that is the ultimate conclusion that we will reach as to what is wrong.

- **What are the necessary and sufficient indicators in the evidence that the problem is of a particular class?**

Answering this question appears to be something that becomes an automatic process when one achieves a particular degree of expertise in the field. However, by working backwards from the conclusions we can determine the signs and symptoms which had been necessary and sufficient to reach that particular conclusion with a particular degree of certainty.

- **What is the evidence?**

For this purpose, evidence is that which is visible, measurable and recordable.

In the example of dampness, this amounts to deterioration of some kind – even if only visual or aesthetic – of part of the building:

- that which is visible will be the presence of stains and/or water in the building;
- that which is measurable and recordable may be the moisture contents and the temperatures of parts of the building and its external and internal environment, together with the manner in which the "visible" and "measurable" changes with time

- The evidence itself needs to be classified, and in a way which will distinguish between the classes of problem. For example, the element of the building on which the problem is present will determine the probability of the various potential conclusions.

There are other items of evidence, but the purpose here is to show, using dampness as an example, how a framework for systematic diagnosis of problems of all kinds might be developed.

In the foregoing it has been convenient to display the elements of the diagnostic process by working from the problem "downward" to the evidence. The application of diagnosis of course requires one to work "upward" from the evidence to the problem (as shown in Figure 3.4).

3.4.2 The deductive approach

Some writers on the subject have equated building pathology with detective work (Fielding 1987). This is because essentially it involves the deductive process as illustrated in Figure 3.5.

According to Addleson (1992), "crime detection, pathology or even forensic medicine provide the precedents for the procedure that should be followed in the diagnosis of building defects". This means that construction

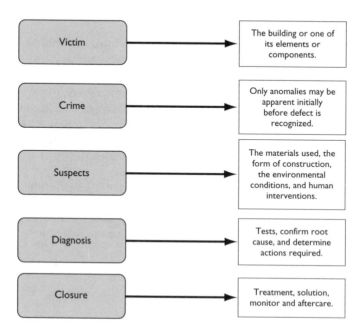

Figure 3.5 The deductive approach (adapted from Addleson 1992).

professionals should bear in mind a number of points when investigating a building defect:

- A defect must not be tackled with any preconceived notions. Avoid jumping to conclusions. It is very easy to assume, for example, that a tidemark on a wall at ground level is caused by so-called true or direct "rising dampness". It is more likely to be caused by bridging (i.e. indirect rising) dampness.
- All similar/common faults may not be ascribable to the same cause. Some building failures may be attributed to more than one causal factor. For example, either sulphate attack or loss of bond from substrate may have caused cracking of external rendering.
- Information that is given or obtained from external sources must be treated with caution. The accuracy and relevancy of the data and how recent it is should be verified. Similarly, data from instruments such as moisture-reading meters should be handled with care, as when used on a carbonaceous or metallic material they will give exaggerated results.
- Several problems can easily be misinterpreted – such as assuming that fungal attack in skirtings and mould growth on the wall surface above has been caused by "rising" dampness.
- Drawings should not be taken as a necessarily correct representation of the actual construction. Changes during construction may result in a detail being different from the as-designed drawing.

Addleson (1992) was right to emphasise that the most important and most difficult part of the defect investigation process is the comparison of symptoms with known behaviour of the materials involved, and the conditions to which they have been exposed and under which they have had to perform. This is where the cause of the defect is deduced.

3.4.3 The "HEIR" approach

This approach, propounded by Rushton (1992), offers a useful summary of the key stages in any systematic building investigation. "H" stands for "history", which means that establishing the chronology and context of the defect/failure is a vital first step. "E" stands for "examination", and represents the reconnaissance and initial site visit stage. "I" stands for "investigation", and represents the detailed inspection accompanied by research and testing as required. Finally, "R" stands for "response", which is detailed in a technical report giving conclusions and recommendations as to remedial work or further actions (Rushton 1992).

3.4.4 Principles for building

The principles *for* building should not be confused with the principles *of* building. The latter are those rules and precautions that relate to building practice. The rules and precautions may (and do) change with experience – the principles applicable to buildings (described by Addleson and Rice 1992, as "Principles for Building") do not.

Examples of some of the principles of building are as follows:

(i) Rules:

- A cavity in a cavity wall should be at least 50 mm wide.
- Mortar or render should not be stronger than the units or background it bonds together or is applied to.
- Damp proof courses should be at least 150 mm above outside ground level.

(ii) Precautions:

- Using cavity battens to avoid mortar droppings.
- Protecting materials such as bricks against the elements before laying.
- Cleaning and preparing surfaces before applying adhesives or bond-dependent materials (such as sealants and paints).

The key aspects of the principles for building are summarised in Figure 3.6.

3.4.5 Three domains model

Cook and Hinks (1992) propose a three-domain model as a systematic approach for analysing building defects. They have classified building defects into the following three broad generic groupings: structural, hygrothermal and environmental. Table 3.6 summarises some aspects of these domains in relation to phases in time.

3.4.6 Three key defect risk areas model

Building defects can be classified into two main groups: typical/common and atypical/unusual (Figure 3.7). This can be described as the third level of decision making in defects analysis. Typically, deterioration of a building's structure and fabric is generally associated with one of three causes: dampness, movement (physical change) and biochemical change (Addleson and Rice 1992). All buildings to a greater or lesser degree are exposed to these three influences or key defect risk areas. Within each of these three causes, however, are numerous agencies or sources such as condensation (i.e. dampness), dry rot (i.e. bio-decay) and subsidence

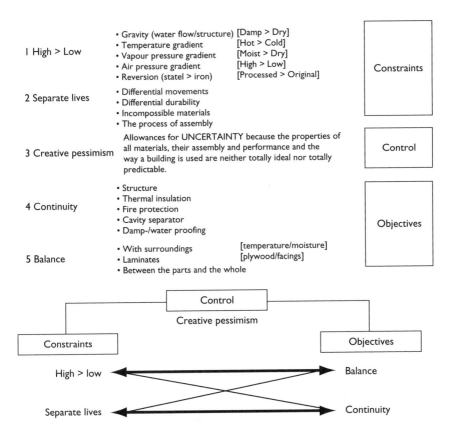

Figure 3.6 Principles for building (Addleson and Rice 1992).

Table 3.6 The three domains for appraising defects (adapted from Cook and Hinks 1992)

| | Domains | | |
	Structural	Hygrothermal	Environmental
Defect Example	Cracking	Rainwater penetration	Poor indoor air quality
Phases			
Initial	Empirical knowledge	Weather resisting	Limited user expectations
Development	New structural forms and materials	Partially sealed buildings	Some buildings unhealthy
Current	Continuing growth in scientific knowledge	Totally sealed buildings	High standards of comfort demanded by users.

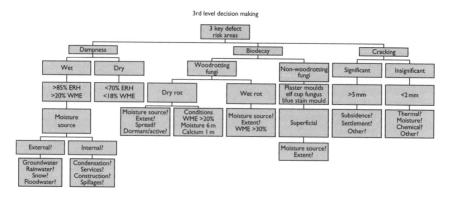

Figure 3.7 Three key defect risk areas model.

(i.e. movement). As pointed out by Addleson (1992), the basic causes of defects in buildings are comparatively few but the agencies (or sources) are many.

Defect diagnosis is made complicated, therefore, given that there are multiple sources (or agencies) under each of these three main "causes". For example, typical sources for the three key defect risk areas are

- *Dampness*: Rain, groundwater, leaking plumbing and air moisture condensation (Burkinshaw and Parrett 2003; Trotman *et al.* 2004).
- *Movement*: Changes in moisture or temperature, loadings, subsidence, settlement, and chemical-induced actions (Bonshor and Bonshor 1996; Dickinson and Thornton 2004).
- *Biochemical*: Chemical incompatibility, moisture-induced problems such as fungal attack, corrosion, sulphate attack (Richardson 2001; Ridout 1999).

It is no surprise that these three key risks are the main defect categories specifically addressed in the RICS's Homebuyers Survey and Valuation pro forma. Generally speaking, if a building were unaffected by these three defects, prima facie its condition could be rated as reasonably satisfactory. However, the building may contain problems such as deleterious materials or other risks such as flooding, illicit alterations, etc. (Noy and Douglas 2005), which could adversely affect its value, potential or safety. One must, therefore, be careful about assuming that such a building is defect-free.

See Chapters 7 to 14 for an examination of the general defects affecting buildings other than these three key risk areas.

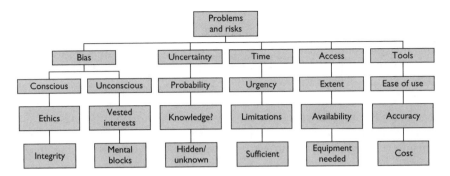

Figure 3.8 Problems and risks in diagnosing defects.

3.5   Problems in defects diagnosis

3.5.1   Main constraints

Investigating building defects is a decision-making process that is constrained or influenced by many factors. Time-pressures, biases, irrationalities and missing or unrecoverable information can easily disturb or inhibit the investigation process (as summarised in Figure 3.8).

The main constraints or problems affecting the efficiency of defects diagnosis are:

- Gaps in investigator's knowledge:
Lack of experience in dealing with the defect being investigated; unknown factors such as original construction; uncertainty over existing construction conditions; level of maintenance; exposure and use conditions; history of anomaly; previous unsuccessful attempts to rectify the problem, etc.

- Investigator overconfident or complacent
These attitudes or attributes can lead to the investigator making rash judgements or jumping to conclusions in overestimating one's competency. Adopting a critical thinking approach will help to avoid such pitfalls (Cottrell 2005).

- Lack of accessibility:
Inaccessible locations in a building, especially void areas or parts of the structure and fabric behind finishings, can make it difficult to inspect the source of a defect.

- Complex interactions:
For example, reactions between damp masonry and timber, resulting in spread of fungal decay like dry rot, which only occurs in the presence of

calcium-bearing materials such as stone, brick and mineral fibre (Bech-Andersen 1995).

- Insufficient time allowed or timing of investigation inappropriate:
Insufficient time, which may cause the investigator to rush the diagnosis, and in appropriate timing of investigation, e.g. investigating a condensation dampness problem in the summer, when it is less prevalent, increase the risk of error.

- Expertise limited:
The investigator may need to call for more specialist input, such as a chemist or materials scientist to analyse the composition of a material using laboratory techniques such as infrared spectroscopy (Douglas and McEwen 1998).

- Deficiencies in data:
Incomplete knowledge or information; reliability of data may be suspect because of inaccurate or poorly calibrated instrument/s used, etc.

- Predictability:
This may be difficult owing to uncertainties, which should be handled in a probabilistic way (CIB 1993). Such an approach can help reduce the incidence of misdiagnosis.

- Red herrings:
Misleading signs and symptoms, such as spillages from cleaning processes or construction moisture being mistaken for "rising dampness". It is easy to reach these conclusions when investigating a building problem, particularly where moisture is involved.

3.5.2 Typical diagnostic errors

A fundamental premise of this book is that a sound and systematic methodology when undertaking an investigation to establish the cause of a defect is required to minimise any risk of misdiagnosis (Addleson 1992). Realistically, however, mistakes or errors can never be eliminated entirely. Imperfections in human behaviour and knowledge as well as economic pressures will continue to undermine the investigator's performance in making correct diagnoses.

Nevertheless, awareness of common errors and pitfalls should help reduce poor diagnostic performance. Building investigators should, in particular, avoid the following typical deficiencies in diagnosis:

- Mistaking rising/bridging dampness for condensation or vice versa.
- Mistaking dry rot for wet rot or vice versa.
- Mistaking cracking induced by thermal expansion for structural movement or vice versa.
- Mistaking condensation for penetrating dampness caused by failure to install a cavity tray at abutment of the flat roof of an extension.

- Assuming that the cause of a dampness problem was only one moisture source.
- Assuming an element where an anomaly has occurred was built according to good practice (e.g. a chimney stack could have been constructed without proper damp proof courses and flashings which caused damp staining in the chimney breast below).
- Overlooking or discounting subtle anomalies of hidden or obscure defects such as

  - warped painted timber, which may be indicative of fungal attack;
  - rust staining through brick joints, which may suggest wall tie failure; and,
  - small/medium-size spots of mould growth on walls, which may be indicative of penetrating dampness through debris bridging a cavity wall.

Burkinshaw and Parrett (2003), for example, point out that "diagnosing damp is as much about understanding the pathology of buildings through the ages as it is a technical and surveying skill". If this is not done the risk of making a misdiagnosis increases. Thus the occurrence of bridging dampness (i.e. indirect rising dampness) can easily give the impression that true rising dampness (i.e. direct rising dampness) is occurring. The careless or unwary investigator may then recommend cavalierly the installation of a retro-fit damp proof course. As highlighted by Howell (1994, 1995, and 1996), such remedial work is more often than not unnecessary.

3.5.3 Sources of diagnostic errors

The sources of diagnostic errors identified by Potter and Perry (2001), whilst primarily related to healthcare, are still pertinent to building pathology. These are summarised in Table 3.7.

3.5.4 Diagnostic judgement

The aim in any diagnosis is to achieve either a true positive or true negative result and avoid either false positive or false negative result. A true positive (TP) result is where a building or element was positively diagnosed as truly having the defect suspected. A false positive (FP) in contrast is where a building or element was positively diagnosed but doesn't actually have the defect suspected.

Similarly, where a building or element that was negatively diagnosed truly doesn't have the suspected defect, it's a true negative (TN). This leaves us

Table 3.7 Sources of errors in diagnosis

Source	Example
Collecting	• Lack of knowledge or skill • Inaccurate data • Missing data • Disorganisation
Interpreting	• Inaccurate interpreting of clues • Failure to consider conflicting clues • Using an insufficient number of clues • Using unreliable or invalid data • Failure to consider cultural influences or development stage
Clustering	• Insufficient cluster of clues • Premature or early closure • Incorrect clustering
Labelling	• Wrong diagnostic label selected • Condition is a collaborative problem • Failure to validate diagnosis with client • Failure to seek guidance

with that which was negatively diagnosed, but really does have the suspected defect – it's a false negative (FN).

Examples of these outcomes in the case of an investigation into the suspected occurrence of condensation are illustrated in Table 3.8.

There is a natural tendency to err on the "safe" side when investigating/inspecting a building and encountering a potential or suspected outbreak of fungal attack or other damp-related problems. The danger of litigation or additional repairs being required is the main driver for this error. It means that more often than not dry rot is diagnosed when fungal attack is suspected in a building. If this were a false positive it could result in unnecessary and often drastic remedial being recommended. In the context of buildings of architectural or historic importance this approach is very damaging. The same problem exists in the investigation of subsidence. There's often the tendency to overstate the problem, which results in excessive or even unnecessary underpinning being recommended.

On the other hand, if the diagnosis were a false positive the wrong type of remedial work could be recommended. For example, wet-rot repairs instead of dry-rot eradication work could be specified. Thus the dry rot attack would more than likely not be eradicated. The risk would be high of a further outbreak within six months to a year of this most dangerous form of fungal attack.

Table 3.8 Diagnostic judgement outcomes

Outcome	Correct diagnosis	Incorrect diagnosis
Judgement +ve (but not necessarily correct)	**True Positive:** Condensation present/occurred • RH% >70% indoors. • Indoor air temperature levels <18°C. • Low surface temperature. • High dew-point temperature. • Mould growth on internal walls. • Sweating on glazing and tiled surfaces.	**False Positive:** Condensation incorrectly suspected • RH occasionally >70%. • Patches of mould on wall. • Lack of background ventilation. • Low level of thermal insulation. • Surface temperature above dew-point.
Judgement −ve (but not necessarily incorrect)	**True Negative:** Condensation absent • RH% <60% indoors. • No mould growth visible. • Air and fabric indoors >16°C. • No signs of sweating on wall or ceiling surfaces (i.e. no apparent surface condensation). • Dew-point temperature well below surface temperature.	**False Negative:** Condensation erroneously ruled out • No obvious signs of mould growth – but behind wardrobes and picture frames. • Cold bridging occurring but not obvious. • No signs of sweating on glazing and walls during the day but at night RH levels high and air temperature levels low, allowing condensation to occur. • Interstitial (i.e. "invisible" condensation).

3.6   Summary

A systematic approach to the investigation of building failures is the surest way of maximising the achievement of successful defects diagnosis and minimising the execution of unnecessary repairs. Loss of professional credibility, more litigation and increasing bad publicity for the construction industry may occur if such an approach is not taken.

Chapter 4

Diagnostic techniques and tools

OVERVIEW

The various scientific and technical methods for assisting diagnosis are addressed in this chapter. It includes destructive as well as non-destructive testing as aids to diagnosis.

4.1 Information and decision aids

Information and decision aids comprise diagnostic charts, matrices, tables and computer-based expert systems (CIB 1993). An example of diagnostic information (for collection)–that is, a schedule of defects – is shown in Appendix B. The following decision aids (for diagnosis) are contained in Appendix C.

- General dampness assessment checklist
- Fabric dampness data checklist
- Moisture content of masonry samples checklist
- Condensation assessment checklist
- Fungal and insect attack checklist
- Crack damage checklist.

Appendix D shows an example of a sample diagnostic report. It follows the diagnostics methodology outlined in Chapter 2.

Other paper-based diagnostic aids are decision trees (as highlighted in Chapter 2), diagnostic trees and fault trees (see Glossary). These are graphical tools that help the investigator to achieve a logical and complete analysis of a given problem and to avoid overlooking possible causes (CIB 1993).

The development of expert or knowledge-based systems as diagnostic aids were reported by the CIB (1993) and Rougier and Lefley (1995). Despite their potential advantages (e.g. accuracy and speed of diagnosis), the drawbacks of expert systems for defects diagnosis (e.g. cost, hardware/software requirements, expertise needed to operate) seemed to have inhibited their widespread use in Building Pathology. Nevertheless, with the ongoing rapid advances made in information technology, it's possible that ad hoc expert systems for defects diagnosis will become more readily available. In the meantime, however, practitioners still need to develop their diagnostic skills and familiarity with the physical tools at their disposal.

4.2 Physical aids

4.2.1 Rationale

Investigators cannot rely on the five senses alone to support a diagnosis. Every building pathologist, therefore, needs to have access to a minimum range of equipment for undertaking even the simplest of investigations. In employing such tools, however, it is imperative that

- The investigator is familiar with and proficient in their use and limitations.
- New equipment is tested and calibrated before use to ensure accuracy.
- Old equipment is checked and recalibrated for accuracy (Houghton-Evans 2005).

The main categories of equipment that can be used in building investigations are summarised below.

4.2.2 Access equipment and tools

- Folding/telescopic ladder (3–4 m reach).
- Manhole-cover keys.
- Screwdriver – for opening up access cover panels.
- Power (hammer) drill – for forming borescope holes
- Hammer and cold chisel – for taking off sample of stone, brick, render or mortar, etc.

4.2.3 Measurement equipment and tools

- Linen/metallic tapes (30 m).
- Spring steel tape (5 m).
- Folding rod (expanded to 2 m).
- Ultrasonic/red laser distance measurer – up to at least 10 m.
- Spirit level and plumb bob – for checking the verticality of walls.
- Theodolite/builders level and staff or laser level and plumb meter.
- Electronic moisture-reading meter (e.g. "Protimeter" *Surveymaster SM*).
- Thermo-hygrometer (e.g. "Protimeter" *Hygromaster*).

- Vernier callipers or tell-tale crack measurer.
- Cover meter/metal detector (see Noy and Douglas 2005).

4.2.4 Protective gear

- Overalls – "Tyvek" or other "breathable" suit material over clothing for working in dirty or contaminated environments.
- Hard-hat (with adjustable visor and headlamp).
- Safety glasses, which can be worn over ordinary spectacles if necessary.
- Face mask or respirator (such as an N95 or N100 mask) – for inspecting dusty enclosures or other environments containing contaminated air.
- Ear muffs – for wearing when being exposed to high-pitch noises from drills, plant, etc.
- Protective boots – nail-proof soles and toe-cap protectors.
- High-visibility over-jacket vest.

4.2.5 Inspection equipment

- High lux torch (e.g. 1 million candle-power minimum).
- Hand mirror with telescopic handle to inspect behind awkward roof spaces, etc.
- Binoculars – for inspecting inaccessible roofs and other elevated positions from ground level.
- Hand lens/magnifying glass for close-up inspections of timber, etc., to check for decay.
- Penknife or bradawl – for poking timber for signs of decay.
- Endoscope (see Noy and Douglas 2005)
- Water-pipe leak listening stick (see Burkinshaw and Parrett 2003).
- Sets of spare bulbs and batteries (for meters, torch, cameras) are useful if not essential.

4.2.6 Recording equipment and tools

- Set of HB-grade pencils.
- All-weather notepad and eraser.
- Set of wax chalk sticks for making markings on suspect points or highlighting important positions.
- Hand-held digital dictation recorder for taking notes on site.
- Set of sealable plastic sample bags with identification label/panel
- Cameras – especially digital camera (with spare memory card).
- Compass to confirm orientation of property.
- Metal detector (see Noy and Douglas 2005)
- Photogrammetry camera – for recording building elevations.
- Rectified photography camera – also for recording building elevations.

4.3   Background to analytical techniques

This section considers the contribution that scientific techniques, such as chemical and laboratory analyses can make in helping to identify the source or make-up of a building defect. The time and money invested is insignificant compared to the consequences of not identifying the correct agency that triggered the defect because of lack of data. An incorrect diagnosis is more likely to lead to the wrong remedy being adopted.

In some cases, building surveyors and other appraisers of defects may need the expertise of specialists to help them solve difficult construction failures. Scientists such as analytical chemists can offer a highly specialised and useful service in this regard. This is particularly so in complex or important investigations. Such inputs, though, are usually only justified or required in cases involving problems in either major new-build projects or large existing buildings. Because of the costs and time involved, they are unlikely to be used in relatively minor or trivial investigations.

Chemical analysis techniques such as spectroscopy are useful as aids to the diagnosis of building defects, especially those of a non-structural kind. Structural building failures, on the other hand, often require analytical techniques at the macro level (i.e. considering an element or the building as a whole) such as radiography or ultrasonics (see under NDT below), for example, to ascertain the extent of voids in concrete elements.

There are also techniques at the micro level (i.e. considering the chemical and physical properties of materials) such as those involving the analysis of samples of concrete and masonry products. Chemical analysis of construction and building materials is usually directed at establishing matters such as

- the cement content of concrete and mortar;
- cement type in concrete/mortar (e.g. high alumina cement);
- concrete or mortar mix proportions;
- type of aggregate and grading;
- the presence of contaminant substances (e.g. organic matter) in a material;
- identification of elements within a material;
- identification of surface deposits, stains, colour variation, foreign bodies, environmental samples (quartz and asbestos);
- identification of preservative chemicals in "treated" timber (see CHN technique below);
- sulphate, sea water and other forms of chemical attack on concrete and masonry products;
- inadequate curing of cementitious and synthetic materials; and,
- the percentage of carbonation with depth in concrete, etc.

Nowadays, scientists employ a number of analytical tools to research and develop the properties and characteristics of new and existing materials. Spectroscopy, X-ray diffraction, and gas/liquid chromatography are amongst the more important techniques (see next section). Some of these techniques can also be used as an aid to diagnosis in the investigation of defects in construction and building materials.

Along with spectroscopy, they can therefore provide investigators with powerful analytical tools to establish the composition of the materials under examination. Although mass spectroscopy, chromatography or X-ray diffraction are not explained here, most of these complementary techniques also have their potential uses in defect analysis. In addition, any one of these techniques is often or routinely used in conjunction with one another.

4.4 Spectroscopy

4.4.1 What is spectroscopy?

Spectroscopy is a specialised branch of the science of optics, which involves the study of the interaction of electromagnetic radiation with matter. It deals with the measurement of light or similar electromagnetic energy after it has been separated into its component wavelengths. The absorption, emission or scattering of light by atoms or molecules can be measured using a sophisticated instrument such as a spectrometer. The results enable scientists to study the properties and constitution of matter at the molecular and sub-molecular levels.

The application and interpretation of spectroscopy requires a sound understanding of chemical analysis. It is not the intention here therefore to give anything more than a cursory explanation of it for non-scientists. In particular, this chapter focuses mainly on one of the most important spectroscopic techniques: infrared spectroscopy.

The various types of spectroscopy are determined by their use of a particular region of the electromagnetic spectrum (Figure 4.1). Infrared spectroscopy, for example, occupies the near mid-range of the spectrum, just above the visible region.

The main spectroscopic techniques are summarised in Table 4.1 below.

Table 4.1 does not include mass spectroscopy. Strictly speaking, because the interaction of molecules with electromagnetic radiation does not occur in a mass spectrometer, it is not a true spectroscopic technique. Mass spectrometers have a mass analyser in place of a frequency analyser, but have many of the features of a spectrometer such as a source of radiation, a sample holder, a detector, and a recorder (Douglas and McEwen 1998).

However, mass X-ray spectroscopy can still be used in the analysis of construction materials. It's particularly applicable for analysing the make-up of mineral-based and other non-organic materials.

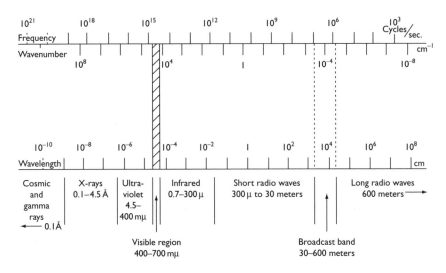

Figure 4.1 The electromagnetic spectrum (Davison and Davison 1967).

4.4.2 How does spectroscopy work?

Science tells us that all materials contain molecules, which in turn are made up of atoms. The atoms are held together in a molecule by links called bonds; these can be regarded as springs with certain vibrational frequencies, which are characteristic of the atom species and the type of bonding. Sharing one or more electron pairs between the atoms forms the chemical bonds in the types of molecules most usefully examined by spectroscopy. The strength of the bond (spring) and the masses of the atoms joined by it dictate the characteristic vibrational frequency of the bond. The reverse is also true; if a certain vibrational frequency can be observed it can be correlated with the bond type and atoms involved or, in more general terms, with the type of chemical group that is present.

Light, being a form of energy, can interact with matter. In the particular case of the infrared region of the electromagnetic spectrum, the energies available are capable of exciting bond vibrations. In doing so, the electromagnetic energy is absorbed by the molecule and converted to vibrational energy – in effect, converted to heat. Spectrometers work by measuring the amount of radiant energy that is absorbed by the molecule at a particular frequency. This is of course the frequency of the bond vibration, which has been excited, and this then can be compared with published correlations between frequency (or more usually wave number) and the chemical group present. The electromagnetic radiation originates from the source and is directed at the samples where some radiation is absorbed and some transmitted. The resultant radiation passes on to a detector, and

Table 4.1 The main types of spectroscopy

Type of Spectroscopy	Comments
Atomic absorption/emission spectroscopy	A form of instrumental/wet chemical analysis. Used for elemental analysis, observes the absorption spectra in the gaseous state.
Electron spin resonance spectroscopy	Used in the study of species with one or more unpaired electrons (radicals).
Electronic (ultraviolet-visible, UV-VIS) spectroscopy	Absorption spectroscopy [100–200 nm (vacuum-UV), 200–800 nm (near-UV and visible)] used to study transitions between atomic and molecular electron energy levels.
Far infrared spectroscopy	Infrared spectroscopy below about $200 \, cm^{-1}$.
Fluorescence spectroscopy (UV-VIS range)	Used to study compounds that fluoresce, phosphoresce or luminesce; fluorescence is the emission energy.
Infrared (IRS) spectroscopy	A form of vibrational spectroscopy; absorptions are usually recorded in the range $200–4000 \, cm^{-1}$. Extremely useful as a "fingerprinting" technique in construction as well as other industries.
Microwave spectroscopy	Absorption spectroscopy used to study the rotational aspects of gas molecules.
Mossbaur spectroscopy	Absorption of y-radiation by certain nuclei (e.g. ^{57}Fe and ^{197}Au); used to study the chemical environment, including oxidation state, of the nuclei.
Nuclear Magnetic Resonance (NMR) spectroscopy	Absorption or emission of radiofrequency radiation used to observe nuclei spin states; a very powerful analytical tool which is used to elucidate molecular structures and study dynamic behaviour in solution and in the solid state.
Photoelectron spectroscopy	Absorption spectroscopy used to study the energies of occupied atomic and molecular orbitals.
Raman spectroscopy	A form of vibrational spectroscopy but with different selection rules from IR spectroscopy; some modes that are IR inactive are Raman active.

the information is then output in the form of a spectrum (Housecroft and Constable 1997).

4.4.3 Different kinds of spectra and their units

The key units are as follows

- Frequency v
  Cycles per second (cps, c/s, or Hertz, Hz)
- Wavelength λ
  Metre (m)
- Wave number

$1/\lambda$ where here λ is in cm, that is the number of waves per centimetre (cm^{-1})

Energy (E) is related to frequency by

$$E = h\upsilon$$

where h is Planck's constant, and frequency is related to wavelength by

$$c = \lambda \upsilon$$

where c is the speed of light. In principle then any of the above units could be used to characterise the energy absorbed by a molecular vibration and different types of spectroscopy tend to use the unit dictated by custom and convenience. Infrared spectroscopy almost always uses the wavenumber, in units of cm^{-1}, which is directly proportional to energy.

4.4.4 Infrared spectroscopy

Infrared spectroscopy (IRS) is a popular technique for analysing the composition of organic materials. It is an excellent "fingerprinting" technique – used to indicate the chemical make-up of a material and help determine the presence or absence of certain compounds.

Infrared spectroscopic analysis was developed in the late 1800s by scientists such as Albert Michelson. It is one of the major and most widely used techniques of analysis in organic chemistry today.

Housecroft and Constable (1997) provide an example of an absorption spectrum consisting of a single absorption, which is schematically represented showing absorbence (the amount of radiant energy absorbed by the molecule) and transmittance (the amount of energy transmitted, i.e. remaining after absorption). In both cases, the spectrum provides data on the measurement of the absorption intensity and frequency. If desired, the frequency of the absorbed radiation can be calculated from the wavenumber, or converted to energy or wavelength.

One of the most important technical developments in recent decades has been the Fourier transform infrared (FTIR) spectrometer. It was named after the French mathematician and mathematical physicist J.B.J Fourier (1768–1830). Fourier developed a method of describing a vibration as a sum of sine or cosine functions. He worked on equations of heat flow, which caused him to develop the Fourier series of the form

$$\Sigma_n \ (a_n \sin)\frac{n\pi x}{k}$$

in order to integrate the basic equation. The Fourier series was adapted later by acousticians and in electrical theory to analyse complex frequency behaviour.

Initially, the FTIR instrument was developed for astronomy in the late 1950s. However, it was not until the 1960s that its use in analytical chemistry became widespread.

The results from an infrared spectrometer are usually presented in terms of transmittance (Housecroft and Constable 1997). FTIR spectrometers have brought about a renaissance in infrared spectroscopy with greater speed of data acquisition and better sensitivity.

4.5   Other analytical techniques

4.5.1   Background

As indicated earlier, there are a variety of other laboratory techniques that can be used as an alternative or supplement to IRS. As with IRS, such techniques are usually destructive, time consuming and expensive. A single, simple laboratory test like one of the chemical analyses described in this chapter, for example, can cost anything between £100 and £500, and may take a day or more to complete. However, in cases where the failure may result in thousands if not tens of thousands of pounds in litigation and repair costs, the cost of such a test may prove very good value for money.

Some of the other analytical techniques that can be used as additional aids in building investigation are as follows

4.5.2   Laser ionisation mass analysis (LIMA)

This is a sensitive "fingerprinting" technique that can help establish known chemical elements in a material sample. LIMA is most effective in analysing the surface of dense materials such as ceramics and metals. It can, however, also be used on organic materials such as timber to help establish if it has been treated with a chemical preservative.

(a) Example

- The LIMA technique was employed by the reviser to confirm whether or not timbers that formed part of a residential conversion scheme to a large property were treated with preservative as specified. The existing sound timbers were left in situ and should have been spray treated with a clear biocide known as "Safeguard Fungicidal Micro Emulsifiable Concentrate" supplied by Safeguard Chemicals Ltd. A second outbreak of dry rot in the premises after the refurbishment suggested that they were not.
- The label on the yellow plastic bottle containing the preservative that was meant to be applied showed that it contained 3-IODO-2-Propynyl-n-Butyl Carbamate. This indicates that iodine is one

of the constituents of the chemical in question. Iodine, which is a well-known antiseptic and sterilisation agent, is a typical ingredient of some modern wood preservatives.

- The label also states that one part of the preservative should be diluted with 24 parts water "to give a final use concentration of 0.2% w/w A.I (2.09g/l)".

(b) Procedure

- A small sample of softwood, approximately 50 mm × 60 mm × 40 mm, was cut from the end of one of the floor joists that had ostensibly been treated. There were no definitely untreated samples of timber available in the property. However, a small piece of untreated wood of similar type was coated with the "Safeguard" preservative to act as a control sample for the LIMA test.
- The LIMA test comprises "a microprobe materials analysis instrument using a focussed laser to ablate and ionise the sample under an ultra high vacuum". Under various intensities the vaporised matter from the lasered surface is then analysed by a mass spectrometer. The resulting spectra of this are then displayed on graph, which shows peaks at the main element positions relating to known masses. Iodine, for example, has a known mass of 127 on the Daltons scale.

(c) Discussion

The first two graphs (Figures 4.2 and 4.3) clearly show the presence of iodine at the known 127 mass position on the Daltons scale. The other peaks in these spectra mainly relate to the elemental composition of the wood. The preservative was applied neatly, so this would explain why the result showed a distinct trace of iodine.

The remaining graphs (Figures 4.4–4.9), which relate to the actual test piece, did not display any peaks at the iodine position. All the other peaks in this sample, however, are consistent with the elemental make-up of the timber.

The iodine should bind with the timber's constituents. Thus, traces of the iodine should have been detected in the timber sample even after the preservative was applied in a diluted state as per the supplier's instructions.

(d) Conclusions

The LIMA test did not detect any iodine in the test piece when compared against the control sample. This suggests that little or no preservative was used to treat the ground floor timbers in the building concerned.

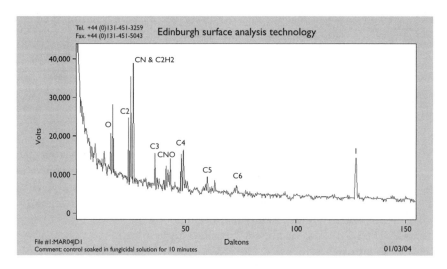

Figure 4.2 Position 1 (Note presence of iodine, identified as "I" at 127, detected between 100 and 150; O = oxygen; C = carbon; C_2 = 2 carbons linked together giving carbon mass $12 \times 2 = 24$; CN carbon and nitrogen combined; C_2H_2 = $2 \times$ carbon + $2 \times$ hydrogen; C_3 = $3 \times$ carbon (linked together three times e.g. C – C – C); CNO = combination of carbon, nitrogen and oxygen; C_4 = $4 \times$ carbon; C5 = $5 \times$ carbon; C_6 = $6 \times$ carbon).

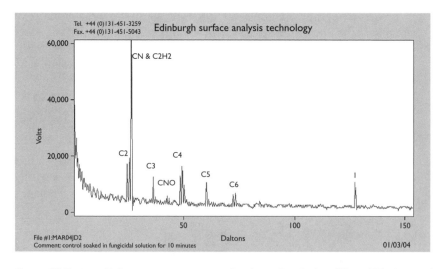

Figure 4.3 Position 2 (again, note presence of iodine, identified as "I" at 127, detected between 100 and 150).

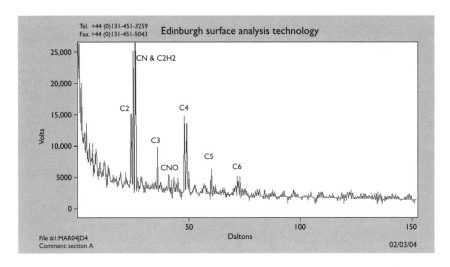

Figure 4.4 Position 1 (note the absence of an iodine peak between 100 and 150 on the Daltons scale).

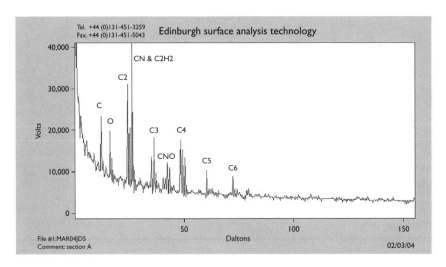

Figure 4.5 Position 2 (note the absence of an iodine peak between 100 and 150 on the Daltons scale).

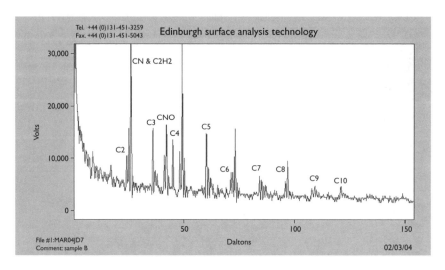

Figure 4.6 Position 1 (note the absence of an iodine peak between 100 and 150 on the Daltons scale).

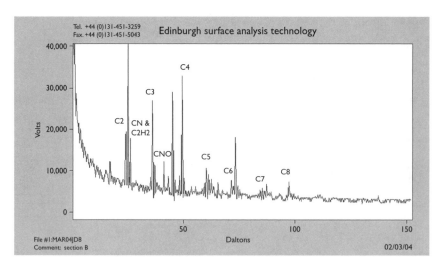

Figure 4.7 Position 2 (again, note the absence of an iodine peak between 100 and 150 on the Daltons scale).

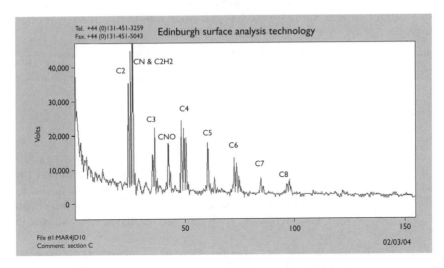

Figure 4.8 Position 1 (note the absence of an iodine peak between 100 and 150 on the Daltons scale).

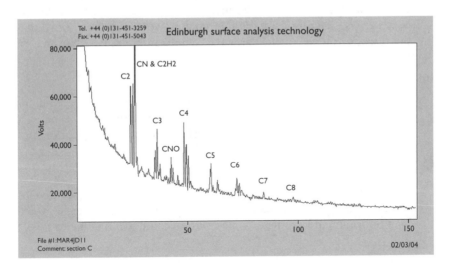

Figure 4.9 Position 2 (again, note the absence of an iodine peak between 100 and 150 on the Daltons scale).

However, the non-appearance of the iodine element (and thus the apparent absence of the preservative) from the test sample cannot be deduced conclusively. It is possible that some of the specified preservative was used to treat the floor timbers but was diluted to such an extent that the low concentrations of the chemical had evaporated. This would also, of course, minimise the level of protection to the treated wood.

4.5.3   Carbon hydrogen nitrogen (CHN) elemental analysis

This technique can be used to analyse building materials such as concrete for organic material that might have contaminated it.

(a) Example

The problem in question was whether or not the surface of the concrete from which a core was taken was contaminated by organic chemicals or materials. Since all organic materials and/or chemicals contain the element carbon, a sample from the core's surface end and a sample (as a control) from the core's underside were analysed for the element carbon using CHN elemental analysis. The results were as shown in Table 4.2 below.

The technique also measures the hydrogen and nitrogen elemental contents, as listed. The latter element, in this case, is *not* diagnostic for organic compounds and is irrelevant. The hydrogen figure could be influenced by differing water contents between the surface and base samples and it too cannot be taken as diagnostic. The carbon figure, however, is diagnostic of organic content, and the data show that the organic content of the surface is significantly greater than that of the base.

(b) Conclusion

The analysis reveals that the surface of the concrete sample contains organic material at a higher concentration than that of the base. It can be inferred that this is due to a surface contaminant of an organic nature.

(c) Caveat

The source or nature of the organic surface contaminant cannot be deduced from this test. Since the concrete had previously been covered by a flooring material the organic residues could well be due to this.

Table 4.2 Results of CHN elemental analysis

Position	Element		
	% Carbon	% Hydrogen	% Nitrogen
Surface	6.5	1.4	0.6
Base	1.0	0.4	0.5

4.5.4 X-ray spectrometry

This method is based on analysis of atomic structure of chemical elements by identifying their mass. It is an effective "fingerprinting" technique for inorganic materials such as cements, ashes, slag, bricks, aggregates and toxic trace elements.

4.5.5 X-ray diffraction

Similar and complementary to X-ray spectrometry, X-ray diffraction is a method of identification of mineral phases present in a material. It is used to identify the following: stains and foreign bodies on materials; asbestos; presence of high alumina cement, chloride attack; inadequate curing; carbonation; and the swelling potential of clay soils.

4.5.6 Gas/liquid chromatography

This technique is used to identify aggressive substances such as oil and fat residues in organic materials. It requires solvent extraction.

4.5.7 Solid phase instrumental analysis

This technique, like X-ray diffraction, can be used for the determination of total sulphur and total carbon content of a material. Companies such as LECO Corporation (www.leco.com/) supply this and other sophisticated analytical instrumentation.

4.5.8 Scanning electron microscopy (SEM)

Scanning electron microscopy is a powerful visual inspection technique for acquiring extremely detailed images of micro-pores and their surrounding structures. It can be used to detect the formation of ettringite crystals in cementitious materials to confirm the occurrence of sulphate attack down to a few microns (μm) and below.

To give an idea of scale, the approximate length of various common small-scale objects are as follows:

- Viruses: c. 30 nanometres (nm) to 300 nm
- Bacteria: c. 0.3 microns (300 nm)–50 microns
- Mould: 4+ microns
- Asbestos (fibres): 3–20 microns
- Pollen: 10–40 microns
- Dust mite faeces: 10–24 microns
- Human hair: 60–80 microns
- House dust mite: c. 300 microns

Acc.V Spot Magn   Det WD ├─────────────┤ 2 μm
20.0 kV 4.8 12800x GSE 7.5 0.4 Torr

Figure 4.10 SEM image of concrete with apparent ettringite formation.

A SEM examination of cementitious materials, for example, can help ascertain the presence of ettringite. Ettringite (calcium sulphoaluminate) is a mineral that has a distinct narrow, needle-like crystal form. It occurs in excess with gypsum (calcium sulphate) when sulphate attack takes place in cementitious materials owing to the presence of tricalcium aluminate (C_3A) in cement (Neville and Brooks 1987). A significant increase in the level of the ettringite content in a sample of concrete compared to control samples, therefore, would indicate that the former has been affected by some sulphate attack.

Abnormal ettringite formation leads to expansion and disruption in a cementitious material. This results in the material being affected by cracking (usually random, hairline in profile) and spalling.

The results of the electron microscope examination at 12800x magnification are illustrated in Figure 4.10. The presence of ettringite needles in the sample can be seen in the black and white photograph. However, care needs to be taken when interpreting such images as the crystal formation may not be solely due to sulphate attack. A chemical test in accordance with BS 1881: Part 124 would be required to confirm the presence of sulphate attack.

4.5.9 Petrography

According to MG Associates (www.mg-assoc.co.uk/serv04.htm):

Petrographic examination opens up the possibility of a much more detailed assessment of concrete quality than can be obtained by any other method. It involves preparing both thin sections and polished plates from samples, usually from cores. The samples are then impregnated with a penetrative resin containing a yellow fluorescent dye and examined under a powerful microscope.

The technique can be used to ascertain

- The composition of the concrete including mix proportions and free water/cement ratio
- The presence and position of reinforcement.
- The extent to which reinforcement is corroded.
- The nature of the external surfaces of the concrete.
- The features and distribution of macro and fine cracks.
- The distribution and size range and type of the aggregate.
- The type and condition of the cement paste.
- Any superficial evidence of deleterious processes affecting the concrete.

4.5.10 Basic chemical analyses of building materials

Most of the following tests require a small sample of material – usually about the size of a matchbox (see also Douglas 1995).

- Acid test to determine the approximate durability of slate, stone, etc.
- Cement content of mortar: BS 4551: 1996
- Cement content of concrete: BS 1881-131: 1998
- Chloride/sulphate content of concrete: BS 1881-124: 1988
- "Protimeter" salts analysis tests of mortar, brick, etc., to determine on site the presence of chlorides and nitrates in a dampness investigation.

4.5.11 Miscellaneous tests

- *Magnesium content*: Acid extract by atomic absorption spectrophotometry.
- *Depth/presence of carbonation*: This can be measured on either a freshly exposed section of a concrete core or an exposed substrate of in situ concrete by spraying it with an indicator solution such as phenolphthalein. When the concrete is alkaline (i.e. say > 9, which means it is providing a chemically protective environment for any reinforcement) it has a pinkish hue, but remains colourless when the concrete is carbonated (i.e. acidic, which means any reinforcement in the concrete is not chemically protected from corrosion).

4.6 Non-destructive testing

4.6.1 Background

A major drawback with all of the preceding techniques is that they usually entail some degree of destruction to the material being tested. This is especially problematic in buildings of historic or architectural importance. The aim therefore when undertaking any such investigation is to cause as little irreversible damage as possible to the structure and fabric of the building. This is where non-destructive testing (NDT) methods can prove extremely useful.

See also Historic Scotland's TAN 23 by GB Geotechnics Ltd (2001) for a more thorough review of the following and other NDT techniques used in building investigations.

4.6.2 Impulse radar

This electromagnetic NDT technique can be used by building pathologists to locate and measure accurately voids, internal cracks, redundant chimney flues and other discontinuities in mass walls and beneath floors. It's a technology that was originally developed for military and geotechnical purposes.

Impulse radar works by the transmission of pulsed radio energy from an antenna held against the surface, and the reflected energy is picked up by another antenna. This allows a profile of the structure below the surface to be analysed, as energy is reflected in varying proportions by inconsistencies within the material. It also provides a useful check if voids have been filled after works.

4.6.3 Infrared thermography

Infrared thermography (IRT) can provide a remarkable amount of information on the structure and condition behind the surface and within the fabric of a building. The technique can be used for the following purposes

- Dampness detection – it can help pinpoint leaks in the fabric of buildings. Figures 4.11 and 4.12 illustrate the use of IRT in leak detection.
- Insulation detection – to check for the presence of cold bridging in the fabric.
- Air leakage detection – it can detect areas where heat losses are occurring in the building envelope.
- Detection of hidden detail (e.g. subsurface pipes, flues, ducts, wall ties, etc.).

- Examination of heating systems, for example, efficiency, damaged insulation, blocked pipes, heat distribution, failed control equipment (steam traps), etc.
- Preventative maintenance – by checking the efficiency of plant and equipment.
- Electrical defect detection.
- Leakage from earth embankment dams (Hart 1991).
- It can be used for locating and assessing:

  - structural timber frames behind render or plaster;
  - structural joints behind render;
  - bond failure and moisture ingress in renders;
  - structural variations in lintels;
  - variations in moisture levels; and
  - termite attack in building timbers (see article on this in the website http://www.flirthermography.com/media/13%20James%202002s. pdf).

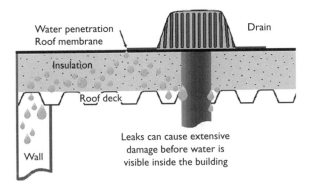

Figure 4.11 Invisible source of water penetration (courtesy of http://www.iranalyzers.com).

Figure 4.12 IRT image showing area where leak is occurring on flat roof (courtesy of http://www.iranalyzers.com).

The equipment used in IRT detects, measures and records minute variations in the infrared radiation, which is emitted by all objects above −273 °C (i.e. absolute zero). Thus, the hotter the object the more intense the radiation it's emitting. Infrared energy is part of the electromagnetic spectrum (see Figure 4.1).

Infrared thermography is a quick method with most works to elevations being carried out from ground level. Images recorded by the camera can be stored on a computer disc and are best printed in colour to aid interpretation.

The weather and building conditions, however, can affect the efficacy of IRT recordings. Rain or very bright sunshine on a light wall or roof being investigated may affect successful assessment. In hot and sunny conditions, in particular, bright surfaces such as shiny metal or solar reflective chippings on flat roofs, or glass, can give results that can be easily misinterpreted (GB Geotechnics Ltd 2001).

Like impulse radar, IRT is a specialist technique that is normally only justified in certain cases – such as for large residential properties or commercial buildings. Specialist companies such as IRT Surveys Ltd (www.irtsurtveys.com) or Infrared Analysers Inc (http://www.iranalyzers.com) can undertake this work for a fee.

4.6.4 Nuclear moisture analysis

This NDT technique is useful for moisture analysis of flat roofs. According to Infrared Analysers Inc. (http://www.iranalyzers.com/nuclearroof.htm):

> Using a principle called neutron moderation, scientists discovered that tiny amounts of radiation can be used to detect hydrogen ions, which in a roof system typically indicate water. Neutrons emitted from an isotopic source collide easily with the highly exposed neutrons of hydrogen, and these collisions slow their travel. By detecting changes in the speed of the emitted neutrons, we can safely identify moisture damage deep within the roof.
>
> To accurately analyze the data from this extremely sensitive nuclear detection technology, we must account for the original hydrogen content of the roofing materials. By utilizing small amounts of invasive testing, hydrogen readings are established for dry materials. After dry material baselines are established, elevated hydrogen counts indicate moisture damage in the roof system.
>
> A nuclear roof moisture survey is typically conducted on a 10′ × 10′ grid pattern across the entire roof surface. On ballasted roofs, the gravel will be temporarily moved aside in order to place the nuclear gauge directly on the roof surface. Readings are taken at each grid intersection to build a hydrogen inventory of the roof system.

Next, we tabulate and analyze the collected data using statistical and visualization software. Using these diagnostic tools, we convert the raw data produced by the nuclear gauge into an AutoCAD® map of moisture penetration inside the roof system.

4.6.5 Microdrilling

The piece of NDT investigative equipment is used for timber – in its natural state and in service. Therefore, it's used by foresters as well as building investigators. It works using a fine drill probe that penetrates up to 200 mm in depth. Linked to a computer, it records faults and variations due to decay and other defects by measuring the speed of penetration. The faster the penetration of the drill, the greater is the decay or unsoundness of the timber.

Extensions of the drill bit allow penetration to depths greater than 200 mm if necessary. It leaves a hole of about 1 mm in diameter, similar to that of the common furniture beetle.

Microdrilling is probably the most accurate NDT method currently available for assessing whether timber can still fulfil its structural role. It's a specialist technique, however, which requires experience to fully assess the results. Moreover, it does not come cheap: a basic microdrill can cost up to £4000 (@ 2006 prices). Seibert Technology Ltd (http://www.sibtec.com/) is a major supplier of this instrument (Noy and Douglas 2005).

4.6.6 Endoscopy

The endoscope is a useful NDT device for inspecting otherwise inaccessible areas in a building: brickwork cavities, underfloor joists, behind panelling, and so on. This type of equipment ranges from the simple borescope, which consists of a light source in a small diameter rigid tube with built-in optics and eyepiece, to more complex and controllable optic fibre systems. The latter can be fitted with a variety of accessories such as steering and digital or video camera attachments as small as 2 mm in diameter.

Endoscopy works by inserting the tube through a 12 mm diameter hole in the building fabric to facilitate inspection of the space behind. Afterward the 12 mm hole left in the brickwork/stonework can be filled with a 1:5 cement sand mortar. In cases of similar boreholes in wood, a timber dowel can be inserted and smoothed off to hide the aperture.

The range of view is about 600 mm. Defects such as fungal growth, defective wall ties, cavity blockages, and infestation within this distance can be identified. Care, however, is required in interpreting the images. Insulation or cobwebs, for example, can be mistaken for dry rot by an inexperienced eye due to scale.

Olympus Industrial (http://www.olympusindustrial.com/) is one of the main suppliers of endoscopes for use in the construction industry. Like microdrills, endoscopes are relatively expensive. A simple prism borescope can cost around £500, whereas a sophisticated optical fibre endoscope with attachments can cost over £2000 (Noy and Douglas 2005).

4.6.7 Moisture measurement

A variety of instruments for measuring moisture are at the building pathologist's disposal. The three most common moisture-reading instruments are the electrical conductance moisture meter, the carbide "Speedy" meter and the hygrometer.

(a) Electrical conductance meter

The "Protimeter Surveymaster SM" is probably the most popular electrical moisture-reading meter used by surveyors in the United Kingdom. It incorporates two modes of operation: search and measure (non-invasive and pin type). Search and measure enable the user to distinguish surface moisture from sub-surface moisture and provide essential information when trying to establish the extent and cause of a moisture problem. The basic instrument (see Noy and Douglas 2005) retails at about £320 at 2007 prices.

(b) "Speedy" calcium carbide meter

This is a useful instrument for ascertaining on site the moisture content of particulate building materials such as mortar, brick, stone, and concrete. The "Speedy" meter and case kit costs around £700 at 2007 prices.

(c) Hygrometer

Condensation analysis requires the use of this instrument to measure air and surface temperatures and the relative humidity of the air. All of these readings are required to establish the dewpoint temperature, which indicates the condensation risk. The "Protimeter Hygromaster" instrument is in the same price range as the "Surveymaster SM".

4.6.8 Schmidt (rebound) hammer

This instrument is intended for non-destructive testing of the quality of concrete in the finished structure (see BS EN 12504-2: 2001). In other words, it gives an approximate indication of the compressive strength of hardened concrete (see also BS 1881-201: 1986). In testing, the "rebound number" is measured, which depends on the strength of mortar close to the surface. Since the strength of mortar as a rule determines the strength of concrete, the rebound number gives an indication of the concrete's quality. The result is the concrete's approximate compressive strength.

4.6.9 Metal detector

This instrument is useful for giving an indication of the presence and depth of reinforcement within concrete or wall ties within brickwork (see Noy and Douglas 2005).

4.6.10 Other NDT techniques

- *X-ray radiography* – used by engineers to assess the composition of hardened concrete to check for fissures and voids.
- *Magnetometry* – used more by archaeologists to obtain an image of sub-surface make-up in floors and walls.
- *Free electromagnetic radiation* – a specialist NDT technique for locating voids and metallic or non-metallic lines buried underground or within walls (GB Geotechnics Ltd 2001).
- *Sniffer dogs* – "rot hounds" are specially trained dogs that can pinpoint the source of a dry-rot outbreak (see www.dryrotdogs.co.uk).

4.7 Summary

A systematic approach to the investigation of building failures is the surest way of maximising the achievement of successful defects diagnosis and minimising the execution of unnecessary repairs. Loss of professional credibility, more litigation and increasing bad publicity for the construction industry may occur if such an approach is not taken.

Analytical techniques are powerful for substantiating a defects diagnosis. They provide an essential source for applying evidence-based practice at the local level especially in the absence of data from randomised control trials.

Chapter 5

Deterioration mechanisms

OVERVIEW

This chapter identifies the main agencies that can trigger or accelerate the deterioration of building elements and materials. It also examines the inter-relationship between deterioration and its adverse consequences such as obsolescence and depreciation.

5.1 Background

5.1.1 Deterioration hypothesis

As indicated earlier, Building Pathology also includes the study and prediction of building life and durability. It is therefore concerned with issues such as deterioration mechanisms, changeability, maintenance cycles, service life and failure rates.

The deterioration hypothesis posits that the condition of a building if left unattended tends to worsen with age. In one sense, it is an example of the Second Law of Thermodynamics, which states that all processes manifest a tendency towards decay and disintegration, with a net increase in what is called the entropy, or state of randomness or disorder, of the system. Thus in the absence of any separate organising force, there is a tendency for things to drift in the direction of greater disorder, or greater entropy.

As highlighted by Douglas (2006), the Second Law of Thermodynamics is applicable to closed (rather than open) energy systems, which tend to fall into disorder. (The First Law of Thermodynamics holds that matter can neither be created nor destroyed – only converted.)

In the context of buildings the activation energy is maintenance and adaptation. These are the two primary interventions that are aimed at combating deterioration and depreciation (see Douglas 2006).

In relation to some building materials, of course, there are a few that seem to be exceptions to this rule. Good quality concrete, for example, tends to harden over time – but this is usually offset by carbonation, which causes

Table 5.1 Seven shearing layers of change (based on Brand 1994)

Layer	Life expectancy or rate of change
Site (including boundary and subsoil conditions)	Permanent
Structure (including foundations and utility connections)	30–300 years
Skin (including roof membrane, cladding and glazing)	20–50 years
Services (including plant and equipment)	7–15 years
Space plan (including circulation)	3–30 years
Stuff (contents)	Daily/monthly
Souls (occupants)	Daily

corrosion of the reinforcement (see Chapter 4). Copper gains a green oxide coating known as verdigris, which acts as an attractive, protective patina. Carbon dioxide performs a similar function on sheet lead by forming a grey patina on its outer surface.

Deleterious physical change, however, is more common. It can occur as a result of the influences of the elements, causing the appearance and the fabric of a building to deteriorate. Typically, transformations in the appearance of a building take the form of soiling of the external envelope (Table 5.1). The accumulation of soot and grime on the facade of old buildings is an extreme example of this influence. Wear and tear as well as user abuse and vandalism also transform the features of a building – usually for the worse.

5.1.2 Shearing layers of building change

Because of the different rates of change of its components, a building is always tearing itself apart. According to Brand (1994), the seven shearing layers of change and their life expectancies are summarised as shown in Table 5.1.

5.2 Principal deterioration mechanisms

5.2.1 Background

Table 5.2 provides a summary of the principal deterioration factors affecting buildings.

When building materials and components are transported, stored on site and used in a structure, they are subjected to the effects of a number of agencies, some of which may influence adversely their durability and performance and, thereby, have a major bearing on the possibility of their

Table 5.2 Summary of deterioration factors

Deterioration factor	Effect
Weathering factors	
• Temperature	
Elevated	Expansion.
Depressed	Frost attack.
Cycles	Freeze/thaw.
• Water	
Solids	Snow, ice.
Liquids	Rain, condensation, standing water.
Vapour	High relative humidity.
• Normal air contaminants	
Oxygen and ozone	Oxidation of metals and plastics.
Carbon dioxide	Carbonation of concrete.
• Other air contaminants	
Gases	Oxides of nitrogen and sulphur.
Mists	Aerosols, salt, acids, and alkalis dissolved in water.
Particulates	Sand, dust and dirt.
• Wind	Erosion, storm damage.
Electromagnetic factors	
• Solar radiation	Heat build-up leading to melting of bituminous, rubberised and polymeric materials.
• Ultraviolet radiation	Photo-oxidation causing bleaching and loss of pigmentation. Thermal degradation.
• Infrared radiation	Radon gas; electromagnetic radiation from
• Nuclear radiation	electricity pylons/overhead power lines and mobile phone masts.
• Lightning	May trigger fire and can cause flash damage particularly to high projections on a building.
Biological factors	
• Fauna	
Woodworm	Timber decay.
Termites	Timber decay.
Birds	Droppings – soiling and disease.
• Fungi	
Wood rotting	Dry/wet rots
Moulds	*Aspergullus, Penicillium*, etc.
Bacteria	Stone decay by aerobic bacteria; degradation of concrete by anaerobic bacteria (e.g. in sewer pipes).
Stress factors	
• Sustained stress	Creep.
• Period stress	Imposed loadings.
• Random stress	Physical action of water, as rain, hail, sleet and snow. Physical action of wind. Combination of physical actions of water and wind. Movement due to other factors, such as settlement or vehicles.

Table 5.2 Continued

Deterioration factor	Effect
Chemical factors	
• Sulphate attack	Cracking of concrete and other cementitious materials.
• Acid rain pollution	Erosion and friability of particulate and metallic materials.
• Alkali attack	On metals such as aluminium.
Incompatibility factors	
• Chemical	Plastics and bitumen products (e.g. causing de-plasticisation of PVC). Dissimilar metals (e.g. causing bimetallic/electrolytic corrosion). Resinous timbers and ferrous fixings (e.g. acidic timber corroding fixings).
• Physical	Porous stonework built on impermeable masonry.
Use Factors	
• Maintenance	Aftercare – inadequate.
• Wear and tear	Abnormal/excessive.
• Abuse/misuse	Neglect or vandalism.

premature failure. The action of the weather, or the external climate, is foremost amongst these. It is often taken to affect only those materials exposed externally. However, the distinction between the external and internal environments in a building is not always clear-cut. There are partially protected areas still open to the weather to some extent. Outside air enters a room through an open window and sunlight is filtered through the window glass. Weather changes in protected or internal environments are usually the same in type but slower in action than those taking place outside. Buildings themselves cause modifications to the weather and to the micro-climate, considerable differences in which may occur in quite short distances. The micro-climate is, indeed, of particular significance and, because of the large number of individual circumstances which can cause its modification, is still an area of much uncertainty and a fruitful one for detailed investigation.

The principal components of the weather include

1. *Radiation* – from the sun and from the rest of the sky; varying in amount, frequency, direction, intensity and spectral composition.
2. *Rain* – varying in direction, droplet size, quantity, intensity, duration, temperature and distribution.
3. *Solidified water* – snow or hail; varying in frequency, direction, shape, size, amount and terminal velocity.
4. *Air and its gaseous constituents* – in particular, water vapour, oxides of sulphur, oxygen and carbon dioxide.

5. *Solid and liquid contaminants of air* – dirt, tar and oil particles, salts; varying in composition, distribution, ease of attachment and detachment; chemical effects. (Salt spray near the coast is an important example.)

Durability and performance are also affected by biological agencies (of which moulds, fungi, bacteria and insects are the most important), ground waters and salts, and manufactured products, for example calcium chloride. It is worth making the point that durability is not an inherent property of a material: different materials have different effective durabilities due in part to each of the following – their physical and chemical properties, the function each has to perform and their position on, or in, a building. In practice, each building material or combination of materials tends to respond differently to the influences outlined above, many of which may be active at any one time. These influences are often inter-related in a complex way and may reinforce or oppose one another in different materials. Thus, coincident strong sunlight and dew have a particularly damaging effect upon paint films. On the other hand, unobstructed sunlight following dew deposited on metal will assist evaporation and so reduce the likelihood of corrosion. One agency alone may exert very different effects, depending on its form or intensity. For example, water in the form of rain, washing over a surface, can retard or prevent mould growth, but moisture in the form of repeated condensation can be highly conducive to its formation. For any particular situation, it is necessary to assess the likely combination of agencies and their effects upon durability and performance, and succeeding chapters seek to make this assessment for the most common building situations. As a basis to considering these more complex inter-relationships, it is, nevertheless, useful to consider separately the agencies mentioned and their general effects.

Particulate materials such as brick and stone are susceptible to damage by a number of non-structural agencies. In typical order of severity, these are

- Salt crystallisation: This is probably the main cause of deterioration of stonework (NMAB 1982).
- Frost attack: Brickwork and stonework exposed to high levels of moisture are especially prone to this form of damage.
- Sulphate attack: Sulphates in the material or in soil in contact with it react with moisture to form larger molecules. This causes expansion and cracking of the material.
- Acids: Sulphuric acid and acids from polluted rainfall can seriously erode stone and brickwork.
- Bacteria: Organic acids may exude bacteria, which can also erode masonry.

5.2.2 Solar radiation

Solar radiation is received at the Earth's surface both directly and as long-wave diffuse sky radiation. The proportion of diffuse sky radiation to total radiation received is considerable and cannot be neglected: indeed, it can exceed direct radiation. Solar radiation is absorbed when it strikes an opaque surface. Most building materials are opaque and their absorptivity (the ratio of the radiation absorbed to the incident radiation received) varies, depending upon the nature and colour of the surface. Black non-metallic surfaces have high absorption. Some values for absorptivity of building materials are shown in Table 5.3. The absorption by some materials of bands of short-wave solar radiation (referred to generally as the ultraviolet) can lead to degradation. Such degradation is confined to organic materials, in particular to plastics, some paints and bituminous-based materials.

5.2.3 Temperature effects

The absorption of solar radiation by surfaces is accompanied by a rise in temperature. Building surfaces can also emit long-wavelength radiation and, in so doing, cool. The drop in temperature can be considerable, particularly on clear nights when radiation from black surfaces, such as asphalt roofs, can cause surface temperatures to fall well below shade air temperature. A rise in temperature leads to an increase in the rate of reactions and can accelerate many degradation processes. (An increase of $10\,^{\circ}C$ doubles the rate of many chemical reactions.) High temperatures, in themselves, also lead to high rates of evaporation and volatilisation. Loss of volatiles from bituminous compositions, some plastics, mastics and sealing compounds can cause shrinkage and brittleness. Evaporation of water from cement mixes can lead to early weakness, poor adhesion and cracking. Phase changes may occur, the best-known example being that which occurs in high-alumina cement, the change being closely associated with a loss in strength. Some building materials, for example bitumen, soften or melt with high temperatures. In contrast, temperatures that are permanently below freezing point

Table 5.3 Absorptivity to solar radiation of some common materials (based on CIBSE 1986)

Material	Absorption co-efficient (clean materials)
Aluminium	0.2
White sand-lime bricks	0.4–0.5
Limestone	0.3–0.5
Concrete tiles	0.65
Dark fletton bricks	0.65
Mortar screed	0.8
Asphalt	0.9

can be highly favourable, which is often not recognised, for they ensure the absence of all cyclic thaw – freeze effects of leaching by rain and of liquid moisture migration. Many degradation processes, too, are slowed down.

Temperature changes cause dimensional changes in materials, particularly when the coefficient of expansion is high as, for example, with aluminium and some plastics. These changes cause stresses which, if not accommodated, can exceed the strength of some materials and cause distortion or rupture. Temperature changes can be quite sudden. Sunlight breaking through a frost-laden fog can heat up a surface very rapidly. Rain falling on a sun-heated surface applies a severe quenching shock. Brittle coatings and joints between dissimilar materials can then undergo the first initial breakdown, leading to subsequent deterioration. In the United Kingdom, air temperatures can change by over 20 °C between night and day, and by over 50 °C between maximum summer temperatures and minimum winter ones. The maximum temperature changes on the surface of building materials and the rates of temperature change are often even greater. Black surfaces which are good absorbers of solar radiation and powerful emitters of lower-temperature radiation can show the maximum changes in temperature, and particularly so, if insulated behind the black surface. The range of temperature changes for such surfaces may be double that of the air-temperature changes mentioned above. The coefficients of thermal expansion of some typical building materials are shown in Table 5.4, together with an indication of the unrestrained movements consequent upon a change in temperature of 50 °C – an annual change which can be expected as a minimum for many materials.

5.2.4 Moisture effects

Moisture in solid, liquid or vapour form can be regarded as the principal agent causing deterioration of buildings (see Rose 2005). It is always present in the atmosphere and, when surface temperatures of materials fall sufficiently, condensation can occur, which may be heavy and prolonged. Even under cover, surfaces can become thoroughly wetted, metals may corrode and be brought into aqueous contact with other metals or materials which may lead to electrolytic attack, and glass may be etched. Conditions are particularly conducive to deterioration when moisture condenses in relatively inaccessible crevices from which subsequent evaporation is slow. Rain, particularly when blown by strong winds, can erode soft materials and, washing over a surface, may remove a part of it in solution. Water has a high heat of vaporisation and is, therefore, slow to evaporate. High precipitation, consequently, implies not only a more complete but a more prolonged wetting of materials. When water freezes in the pores of materials, such as brick, stone and concrete, stresses are produced that may cause spalling of the surface, general cracking or disintegration. Water frozen in the form of hail can

Table 5.4 Thermal expansion of some common building materials

Material	Approximate coefficient of linear expansion per °C ($\times 10^{-6}$)	Unrestrained movement for 50°C change (mm/m)
Brick and tiles (fired clay)	5–6	0.25–0.30
Limestone	6–9	0.30–0.45
Glass	7–8	0.35–0.40
Marble	8	0.40
Slates	8	0.40
Granite	8–10	0.40–0.50
Asbestos cement	9–12	0.45–0.60
Concrete and mortars	9–13	0.45–0.65
Mild steel	11	0.55
Bricks (sand-lime)	13–15	0.65–0.75
Stainless steel (austenitic)	17	0.85
Copper	17	0.85
Glass reinforced polyester (GRP)	20	1.00
Aluminium	24	1.20
Lead	29	1.45
Zinc (pure)	31	1.55
PVC (rigid)	50	2.50
PVC (plasticised)	70	3.50
Polycarbonate	70	3.50

cause pitting of some surfaces and, as snow, has to be allowed for in structural design. Changes in relative humidity can lead to dimensional change in materials, with deformation, crazing or cracking. Prolonged low humidities can cause the dehydration of gypsum products (though such humidities do not occur naturally in the UK); prolonged high humidities aid fungal growth and the subsequent decay of organic materials. Moisture also stimulates biological activity and acts as a medium or catalyst through which reactions occur which could not otherwise take place. Because of its major role in causing or assisting failures in materials, components and structures, moisture is dealt with separately and at greater length in Chapter 7.

5.2.5 Biological agencies

Attacks by fungi and insects are principally upon timber (Figure 5.1), though other materials, generally organic, can be affected. In recent years, a good deal of timber decay has been caused by the several varieties of fungi which are covered by the term "wet rot". Wet-rot fungi require markedly damp conditions to germinate and a continuing source of moisture for their existence. Sapwood at a moisture content of around 30 per cent and a temperature around 20°C provides an ideal abode. Timber with a moisture content below 20 per cent is not endangered. Once the source of moisture

Figure 5.1 Rot in roof timbers.

is removed, the fungi will die and do not have the ability to spread to dry timber, or to penetrate plaster and brickwork, as does the fungus *Serpula lacrymans* (dry rot). Spores of the dry-rot fungus are generally present in the air and, given the right conditions, will germinate. Though they are too small to be seen individually by the unaided eye they can be spread by moving air in a building to settle as a rust-coloured dust. The conditions favoured are dark stagnant ones with timber of moisture content above 20 per cent, temperatures around 22 °C, and the presence of calcium next to the affected area (Douglas and Singh 1995). Poorly ventilated sub-floor areas and situations where timber is in prolonged contact with damp materials are those where the danger of attack is greatest. The fungus grows most readily on unsaturated wood which has a moisture content of 30 per cent or more. Once germination has occurred filaments called hyphae spread over the surface to form whitish fluffy growths or sheets known as the mycelium. The hyphae can penetrate cracks in materials such as plaster, brick and block, which in themselves do not provide nourishment, in search of further wood. The growth of the fungus over these inorganic materials is made possible by strands, much thicker than hyphae, which are formed behind the latter and convey food and moisture from the damp wood where the attack began to these hyphae. If further wood is then reached it can be attacked even if not damp for moisture is conveyed to the cellulose in the drier wood by the strands, though there is some evidence that the fungus is not efficient at conducting moisture from a wet to a dry location (BRE

Digest 299). The affected wood loses its cellulose, its strength and its weight and cracks and shrinks often into cubic pieces. Dark, damp and stagnant conditions are also favoured by the various moulds which can, under those circumstances, damage decorations, furnishings and fabrics, and have been the cause of many complaints in local-authority housing subject to acute condensation problems (Garratt and Nowak 1991).

The common furniture beetle (*Anobium punctatum*) is responsible for most of the damage caused to timber by insect attack in the United Kingdom. The eggs are laid in cracks and joints, and when the larvae hatch they bore into the timber. They remain in the timber for around three years, growing all the while, until they reach up to 6 mm long. During this period, the insects tunnel to and fro within the timber, thus weakening it. When fully grown, they emerge from the wood as beetles, leaving circular holes approximately 1.5 mm in diameter (Ridout 1999).

Plywood made from birch and bonded with natural glues is particularly susceptible to attack. More serious damage, though fortunately less widespread, can be caused by the house longhorn beetle (*Hylotrupes bajulus*). The beetle is generally around 10–20 mm long and can cause serious damage to the sapwood of softwoods, particularly to roof timbers. Hardwoods are not attacked. Unfortunately, the larval stage, during which the grub itself can grow to over 30 mm, can last up to 10 years, though 5–6 years is more common in the United Kingdom. During that time, extensive tunnelling and damage can occur. The exit holes are oval, not circular as with *A. punctatum*, widely spaced and up to 10 mm wide. Timber which shows external signs of attack may well be very seriously weakened and so need total removal and destruction. Fortunately, attack by house longhorn beetle in the United Kingdom seems to be confined, so far, mainly to areas in Berkshire, Hampshire and Surrey, and the "Approved Document" to support Regulation 7 of the Building Regulations 2000 requires the special treatment of softwood against the beetle in specified areas within those counties. Damage to hardwoods, usually oak, may also be caused by deathwatch beetle (*Xestobium rufovillosum*). The beetle, however, attacks only those woods which have already been weakened by fungal attack. It is a problem generally confined to old, large structures.

5.2.6 Gaseous constituents and pollutants of air

Sulphur dioxide is generated by the burning of fuel, and concentrations in the atmosphere are greatest in large industrial areas. Great improvements have been made in recent years in preventing harmful emission of sulphur gases – concentrations in London are now around one-third of what they were just after the Second World War. Sulphur gases dissolved in rainwater, nevertheless, still rank as the most aggressive gaseous pollutant and can assist the corrosion of some metals and cause some stone to blister and to spall.

Carbon dioxide is a normal atmospheric gas of little general importance but, dissolved in rainwater, it forms a weak acid capable of slowly eroding limestone and weakening calcareous sandstones by solution of the bonding calcite. It will slowly carbonate lime formed during the hydration of cement. A change in physical properties may result. Thus, asbestos cement becomes more brittle and a lime bloom may appear on concrete products generally. The extent of carbonation in concrete can have a marked influence on the corrosion rate of reinforcement and this is considered in more detail later. As far as is known, nitrogen and the inert gases in the air do not affect building materials. The oxygen content does not vary sufficiently to cause a difference in its oxidative effect on those building materials (such as the organic ones) which are slowly degraded by it. Ozone is rather more variable in distribution. Though present only in traces, it plays the dominant part in the degradation of rubber, particularly when it is stressed, and can be presumed also to influence the degradation of mastics, bituminous compositions, paints and plastics. However, its effects are not of major importance.

5.2.7 Solid contaminants

Dirt and general grime in the atmosphere consist of inorganic dust particles, together with unburnt particles of fuel, bound together by a form of oil or other organic matter derived from fuel, including that from road vehicles. The dirt also contains some soluble salts. Such dirt is deposited on buildings and causes an adverse effect on appearance. There is some evidence, however, that more harmful effects may result, such as an increase in the corrosion rate of metal surfaces and the deterioration of some stone surfaces. Near the coast, the concentration of salt derived from sea-spray is high and the corrosive effects on metals can be severe. Salt particles are deliquescent and absorb moisture from the atmosphere, forming strong solutions, and they are liable to cause severe corrosion if they lodge in crevices. In the close vicinity of particular industries, a variety of gaseous and solid contaminants may affect buildings but, in a national context, the effects are relatively unimportant.

5.2.8 Ground salts and waters

Salts, present in the ground, can rise in solution and by capillarity into porous materials with which they are in contact. Upon subsequent evaporation of the solvent (water), salts may be deposited within the pores or upon the surface of the material, the exact site of deposition depending upon the pore structure and the rate of drying. Usually, little more than efflorescence occurs, often ephemeral and of little consequence. More seriously, if magnesium sulphate is present, disintegration of renderings and masonry surfaces can occur, though failures appear to be infrequent. Peaty moorland ground waters are acidic and can cause concrete in contact with them to lose strength and to erode, the effect increasing with prolonged contact.

5.2.9 Manufactured Products

Manufactured materials may be used as additions to building materials, or as treatments for them, and may have an adverse effect on durability and performance, if not used with care and understanding. Calcium chloride is such a material and is commonly used as an accelerator of the hydration and development of the early strength of cement-based products, for example mortar, concrete and wood wool. Its use has commercial advantages, in that it facilitates the early demoulding of precast elements and their removal from the factory production floor to storage and, also, the early striking of formwork from in situ concrete. However, it has a corrosive effect on metals and severe damage has been caused to reinforced concretes and to pre-stressed concrete through its use. This problem is considered in more detail in Chapter 10.

Inorganic salts may be used as fire retardants and as preservatives in wood. While these have no adverse effects on the wood itself, they can, under some environmental conditions, assist the corrosion of metal fasteners used in timber construction.

5.2.10 Juxtaposition of materials and components

The use of some building materials in proximity to each other can lead to weathering effects, with consequent maintenance expenditure. Staining is probably the most familiar effect and can be caused by design details, resulting in dirt deposition in some places and its removal in others – unsightly streaks on the building are often the result. Products of metal corrosion can leave rust stains on concrete. Efflorescence from brick-work gives a particularly common stain.

Transmission of a soluble component from one part of a building to another, however, may cause not only staining but more serious trouble. Thus, sulphates from bricks can cause breakdown of mortar, corrosion of metals or disintegration of stone. Most timbers are acidic, particularly the heartwood of oak and of Douglas fir, due to the presence of volatile acetic acid, and can corrode some metals. Western red cedar shingles, through their presence in the wood of water-soluble organic derivatives, and lichens on roofs can both attack metal rainwater goods and flashings. The external use of aluminium, copper, phosphor-bronze and other non-ferrous metals, and of stainless steel, needs great care because of the often unforeseen opportunities for electro-chemical corrosion. Washings from limestone onto some sandstone plinths and mouldings can cause deterioration of the latter. It should be emphasised that different materials which, in isolation, may possess adequate durability can deteriorate markedly when brought into aqueous contact with one another. In general, such dangers have been well documented but, with the introduction of newer materials into construction, less well known problems may arise. For example, polystyrene in contact with PVC can lead to loss of plasticiser from the latter, causing it to become brittle and the polystyrene to "shrink" away.

Due to their differences in composition and structure, sandstones can perform very differently in masonry constructions. Thus, sandstones with a higher porosity will absorb water whilst those with a greater clay content will shed water.

If the masonry construction comprises homogenous sandstone, the issue of incompatibility of course should apply. However, where a building has been constructed with a porous stone, difficulties will subsequently arise if any indent repair material has a greater clay content. The latter will simply shed water on to the stones below which have a greater capacity to absorb water. This "double soaking" leads to an acceleration of the decay mechanisms that may well result in distress and even failure of these neighbouring stones.

Thus one repair can induce another – another example of the reverse objective outcome syndrome.

5.3   Degrees of deterioration

5.3.1   Introduction

Physical deterioration manifests itself in a variety of ways. There is, in other words, a range of deterioration of a building (Table 5.5).

In general terms, expressions such as "dilapidated", "crumbling", "wrecked" or "ruinous" typically describe, in order of severity, the extent

Table 5.5 Spectrum of deterioration

Types of deterioration	Range
• Staining or soiling of appearance	Superficial Deterioration
• Degradation of appearance or finish	
• Dimensional changes in element or component	
• Distortion	
• Friability	
• Breakage	
• Disintegration	Irreversible Deterioration

Table 5.6 The spectrum of building condition (Douglas 2006)

Rating	Bad	Poor	Fair	Good
Category	Ruinous	Dilapidated	Wind and water-tight	Optimal
Description	Only some walls left. Little or no roof structure remaining. No windows and doors left.	Extensive defects to the structure and fabric. Crumbling fabric nearing dereliction and redundancy.	Showing signs of wear and tear. Neglected and approaching being run down.	Reasonable state of repair. No major defects. Satisfies most user requirements.
Response	Restoration. Conservation. Conversion to other use.	Renovation. Conversion to same or modified use.	Modernisation. Refurbishment Rehabilitation. Alteration. Extension.	Minor improvements. Some alterations. Extension.

of deterioration. Table 5.6 summarises these and their effects on a building's condition.

The focus of this book is primarily on physical deterioration. However, adverse building change does not only occur in physical terms. The other main types of deterioration are

- Economic: Depreciation or reduction in capital and rental values of the property.

- Legal: Increased non-compliance with statutory requirements.

- Environmental: Increased pollution and blight leading to an unsustainable building.

- Functional: Partial or full cessation of use leading to redundancy of the deteriorated building.

- Social: Reduction in or loss of amenity owing to obsolescence or redundancy.

5.3.2 Changes as symptoms of deterioration

Changes in a variety of physical ways, manifesting via the five human senses, are usually the first indication of deterioration. These are summarised as follows:

- Changes in smell: Mushroomy odour is indicative of dry rot
                          or dampness; noxious or foul odour
                          indicative of a defective drain.

- Changes in taste: Bitter taste in drinking water may be
                          indicative of contamination of the supply.

- Changes in sound: Hollow or dull sound when tapping
                          structural timbers may be indicative
                          of decay.

- Changes in touch: Friability of mortar or stone is indicative of
                          either degradation or poor quality material.

- Changes in appearance: Discolouration of water is indicative of its
                          contamination. See also next section.

5.3.3 Changes in appearance

Changes in the appearance of a building component, element or material can
often signal trouble. The main examples of such changes are summarised
in Table 5.7.

Table 5.7 Typical changes in appearance and their causes

Change	Examples	Possible causes
Shape	Bowing	Wall-tie failure or lack of lateral restraint
	Leaning	Subsidence or settlement
	Distortion	Structural movement
	Deflection	Overloading
	Erosion	Weathering
	Enlargement	Moisture/thermal induced movement
	Misalignment	General movement
	Reduction	Thermal movement
	Warping	Shrinkage
Colour	Bleaching	Ultra-violet degradation
	Lightening	Ageing
	Darkening	Fire/smoke damage, water or oil staining
	Discolouration	Fungal attack or chemical reaction
	Soiling	User abuse
Texture	Cracking	Sulphate attack or shrinkage cracking
	Bubbling	Trapped moisture under coatings
	Bossing	Debonding of coating
	Pock-marking	Acid rain
	Roughness	Weathering
	Smoothness	Wear and tear
	Organic growths	Moisture build-up and lack of maintenance

5.4 Staining

5.4.1 Introduction

Staining and soiling account for the most obvious or conspicuous forms of changes in a building's appearance. The main forms of defacement that affect property are listed in Table 5.8.

5.4.2 Biofouling

"Biofouling" is a term used to describe the defacement of external and internal surfaces by organic contaminants. It can easily disfigure an otherwise attractive surface.

Microorganisms such as fungi and moulds usually cause organic staining on wall, ceiling and floor surfaces. These can have adverse health effects (e.g. mainly respiratory ailments such as asthma, rhinitis, etc.) as well as aesthetic implications (Douglas 2006).

5.4.3 Pattern staining

According to Diamant (1977), if a wall or ceiling is evenly insulated throughout, the dust tends to be deposited evenly. One of the most annoying aspects of the existence of thermal bridges or irregular insulation of walls and ceilings is the formation of dirt patterns. The reason for this is that air

Table 5.8 Types of staining on buildings

	Staining	
Type	External	Internal
Biological	• Algae • Lichen • Mould • Animal excrement	• Mould • Cobwebs • Slime • Human soiling
Physical	• Soot • Grime • Dampness (seismic/rainwater staining) • Dampness (general) • Efflorescence • Graffiti • Oil • Rust • Water-soluble chemicals such as sulphates causing yellow stains on stonework	• Tobacco • Grime • Grease • Dampness (tide mark) • Pattern (thermal) • Hygroscopic salts • Graffiti • Oil • Rust • Water-soluble chemicals such as sulphates causing yellow stains on stonework

molecules (Brownian movement) are constantly bombarding suspended dirt particles. As hotter air molecules move faster than colder ones, there is a concentration of dirt particles over the colder areas of the wall and ceiling, since they are exposed to different impact momenta in different directions. The particles move slower over cold sections and therefore tend to deposit on these.

When the particles finally settle they naturally deposit themselves upon the wall and ceiling areas with the lowest surface temperatures. Thus it can be considered that the distribution of pattern staining is a direct measure of the surface temperature of the wall and ceiling and consequently of the thermal insulation properties of the materials beneath. In the case of an uninsulated plaster boarding and rafter ceiling, lighter lines trace the position of the rafters. If the space between the rafters is insulated by means of glass fibre, while the timber rafters are not, the pattern is often reversed, although it is not as pronounced due to the fact that the difference in temperature between the surfaces affected is reduced. In the case of walls, the maximum amount of pattern staining is usually found near the ceiling and the floor due to the thermal bridges formed by flooring connections. Pattern staining is also found near corners, but the actual corner itself is usually much lighter than its surroundings as the corner is generally better insulated than the wall surfaces on either side.

A further manifestation of pattern staining is the blackening of walls above heating elements, such as hot water radiators or electric fires, and the blackening of patches around electric light bulbs. In these cases, it is difficult to prevent the staining except by the provision of shields to stop the dust particles from alighting on wall and ceiling.

5.4.4 Removing staining and soiling

Generally, taking care to avoid uneven insulation and thermal bridges can prevent pattern staining. For example, the use of about 200–300 mm thick layer of glass fibre insulation matting over the ceiling and rafters almost completely cures dirt patterns there.

For other forms of soiling see BRE Good Repair Guide 27 (Cleaning external walls of buildings: Part 1 – cleaning methods; Part 2 – removing dirt and stains). In addition, Digest 370 (1992) *Control of lichens, moulds and similar growths* should be consulted for dealing with biofouling problems.

5.5 Relationship between deterioration, obsolescence and depreciation

Deterioration is inevitable as part of the ageing process. It is mainly a function of time and use, but can be controlled to a certain degree through maintenance and adaptation (Douglas 2006). As indicated by Ashworth

(1999), deterioration usually has an adverse effect on the performance of a building.

Obsolescence on the other hand is more difficult to predict and control (see Douglas 2006). In a general sense, obsolescence refers to an object's usefulness over time. It indicates the tendency of objects and operations to become out-of-date, outmoded or old-fashioned. It is the transition towards the state of being obsolete, or useless. In short, obsolete means antiquated, disused or discarded.

Depreciation is the economic or financial consequence of deterioration and obsolescence. Maintenance and adaptation, however, can help to combat this (Douglas 2006).

5.6 Summary

Deterioration of a building and its components starts during construction process. Interventions such as maintenance and refurbishment can help stave off if not arrest deterioration. However, sooner or later many of the components and elements of a building cannot resist natural as well as abnormal wear and tear. This is where durability and life service assessment come in and it is to these we turn in Chapter 6.

Durability and service life assessment

OVERVIEW

This chapter examines the main influences that affect the durability of building materials, components and elements. The durability characteristics of the main materials used in buildings are also addressed. The chapter concludes with an introduction to the assessment of service life of buildings and their products.

6.1 Durability of materials

6.1.1 Introduction

Simply put, durability is a measure of a building's ability to resist deterioration (see Glossary). From the day it is completed, the durability and performance of a building's original structure and fabric are eroded or undermined by a variety of agencies (e.g. see BS 7543). These can be classified into two broad groups: environmental and functional (Douglas 2006).

- Environmental influences (climatic agencies):
These particularly affect the exterior of the building, but ultimately impinge on its interior. For example, roof leaks can cause considerable damage and inconvenience within a building. Rainwater can easily ruin stock and equipment as well as damage furniture and finishings (Addleson 1992).

- Functional influences (user activities):
These primarily affect the interior of the building, but can impinge on its exterior. Lack of maintenance, for example, will inevitably have an adverse effect on a property's external condition (Wood 2003). This is most noticeable when vegetation grows profusely in uncleaned roof gutters. The gutters may clog after a few months and cause them to overflow, resulting in water saturating the wall and organic staining on the face of

the building. This can eventually lead to fungal attack of vulnerable timbers, nearby, embedded in the affected masonry (Palfreyman and Urquhart 2002).

The next section will consider the durability of the most common materials used in buildings.

6.1.2 Asbestos cement

Asbestos cement is made from asbestos fibres, ordinary Portland cement and water. The cement hydrates and sets around and between the asbestos fibres, which act as reinforcement. The fibres are based on the mineral chrysolite and have an average length of around 5 mm. The relative proportions of asbestos fibre and cement used vary with the particular product to be made: for roof sheets and slates the proportion of asbestos is around 10 per cent by weight of the dry materials. Admixtures which may also be used in small quantities are principally colouring pigments and cellulose fibre.

The coefficient of thermal expansion of the material is small (see Table 5.4). Changes in moisture content cause far greater movement (Table 6.1). When

Table 6.1 Wetting/drying movements (approximately reversible)

Building material	Approximate movement (%)
Asphalt, bitumen	Nil or negligible
Glass	Nil or negligible
Granite	Nil or negligible
Metals	Nil or negligible
Plastics	Nil or negligible
Clay bricks	0.02
Calcium silicate bricks	0.02–0.04
Concrete bricks	0.04–0.06
Dense concrete	0.03–0.04
Structural lightweight aggregate concrete	0.03–0.08
Concrete with shrinkable aggregates	0.05–0.09
Limestone	0.06–0.08
Sandstone	0.06–0.08
European spruce, Baltic whitewood	1.5* 0.7[t]
European larch	1.7* 0.8[t]
Douglas fir, Oregon pine	1.5* 1.2[t]
Western hemlock	1.9* 0.9[t]
Scots pine, Baltic redwood	2.2* 1.0[t]
English oak	2.5* 1.5[t]

*Tangential
[t] Radial

dry asbestos-cement slates are wetted, they can expand by between 0.1 and 0.3 per cent of their initial dry length: the older the slates, the smaller are their movement. Carbon dioxide, always present in the atmosphere, carbonates the lime formed by the hydration of the Portland cement and this carbonation causes shrinkage of the exposed asbestos cement. Whether the effects of moisture causing expansion, or carbonation causing shrinkage, will predominate is dependent largely upon the age of the asbestos cement and the number of wetting and drying cycles to which it has been subjected already. The greater these are, the greater will become the predominance of the effects of carbonation and, eventually, normal exposure will result in overall shrinkage. Asbestos-cement products are highly resistant to atmospheric pollution, and it is mainly physical changes which make them more prone to damage (DOE 1986). Sheets harden with age through hydration of the cement matrix, which can continue for a long while. Atmospheric carbonation decreases impact strength which, over a long period, can fall to half its initial value. The long-term effect of natural weathering, therefore, is for sheets to become more brittle, and more liable to damage by impact. Both old and new asbestos cement can be damaged by rough handling during transport and erection. When the two faces of a sheet are subjected to widely differing rates of atmospheric carbonation, differential shrinkage occurs. Some instances are known where this has been sufficient to cause cracking; for example, where the external face has been painted or surface-coated, which retards carbonation, while the internal face has been left unpainted. Carbonation shrinkage of asbestos cement, together with the tendency to expand through cement hydration and moisture effects, can cause warping, which is enhanced by any differential movement between the two surfaces. When the external face of asbestos-cement sheets or slates is treated with a surface coating, the internal face should, at least, be primed. This will help to reduce differential moisture movement and carbonation shrinkage between the two faces. Cracking and warping of sheets can occur even when they are correctly fixed to sound supports, unless these precautions are taken.

Asbestos cement is not susceptible to attack by insects but both algal and fungal growths are common on the normal unpainted product. The surfaces become slightly softened, with resultant darkening and discolouration, which is unpleasant aesthetically and, with roofing and cladding sheets, may be undesirable because of the increased ability to absorb solar radiation.

Because of the disabling lung disease asbestosis associated with the use of asbestos and its products, given much prominence in a television programme in 1978, manufacturers of asbestos-cement products have moved rapidly towards the production of non-asbestos substitutes (Asbestos Working Group 2003). The trend has continued but asbestos-cement products, in particular corrugated cladding and roofing sheets and roofing slates, are still being used in many parts of the developing world.

6.1.3 Asphalt and bitumen

A range of products, referred to broadly as bituminous or asphaltic, is used in the building industry, and the terminology can be confusing. The following definitions are based on those contained in BS 6577:

Bitumen – a viscous liquid, or a solid, consisting essentially of hydro-carbons and their derivatives, which is soluble in carbon disulphide. It is substantially non-volatile and softens gradually when heated. It is black or brown in colour, and possesses waterproofing and adhesive properties. It is obtained by refinery processes from petroleum and is also found as a natural deposit or as a component of naturally occurring asphalt, in which it is associated with mineral matter.

Asphalt – a mixture of bitumen with a substantial proportion of inert mineral matter. Asphalt may occur in nature as natural rock asphalt, which is a consolidated calcareous rock impregnated with bitumen exclusively by a natural process. It may also occur as lake asphalt (Trinidad), where it is in a condition of flow or fluidity. Mastic asphalt, which is the asphalt used in building, is a type of asphalt in which the mineral matter is suitably graded in size. Exposure to sunlight over a substantial period causes mastic asphalt and bitumen to harden and to shrink, the shrinking often resulting in slight surface crazing. This effect, which is due to shortwave solar radiation, is assisted by atmospheric oxidation. Being black or dark grey in colour, its absorption of solar radiation is high (Table 6.2), and the high temperatures reached as a consequence can, if the material is not of the right grade, cause softening and flow. Even when correctly graded materials are used, thermal expansion and contraction can be high, through the wide temperature changes resulting from high absorptivity and high emissivity. When such changes are slow, these movements can be accommodated without internal damage but rapid movements through sharp changes in temperature can cause cracking, particularly in cold weather. Under mechanical

Table 6.2 Service life determinants

Influences	Material/component dependent	Material/component independent
Direct	Quality of product Level of protection Initial performance	Exposure Location Wear and tear
Indirect	Original workmanship Standard of maintenance Repairability	Maintenance policy Maintenance level Environmental influences

stress, bitumen products can flow. Asphalt and bitumen are not affected by biological agencies or by pollution but contact with oil can be damaging. Moisture has no direct adverse effect – the materials are used essentially for waterproofing – but moisture vapour pressure can cause blistering. This problem and others affecting the use of asphalt and bitumen for roofing are dealt with later.

6.1.4   Bricks and tiles

Most bricks used are of burnt clay or calcium silicate, the former being classified in BS 3921 and the latter in BS 187. Clay bricks suitable for general building purposes are described as 'common'. Facing bricks are specially made or selected for their attractive appearance. Engineering bricks are strong bricks of low permeability. Thermal expansion of clay bricks is greatly exceeded by an irreversible moisture movement which occurs when newly fired bricks absorb moisture. Typical expansion of an individual brick can be around 0.1 per cent and even close to 0.2 per cent, in some cases; around half of that movement, however, will occur in the first week after manufacture. Such an expansion can cause problems, particularly when associated with opposing movements in structures, and this is considered in more detail further on, where cladding is dealt with. A good deal of potential difficulty can be avoided by not using bricks fresh from the kiln: brickwork, as opposed to individual bricks, suffers less expansion – in general, just over half the amount. A normal reversible wetting and drying movement of some 0.02 per cent occurs in addition to this irreversible expansion.

Clay bricks, which have a water absorption of less than 7 per cent by weight, have a high resistance to damage by frost. It should not be assumed, however, that bricks with a high water absorption have a low resistance, for resistance depends not only on total porosity but also on pore structure and, in particular, upon the proportion of fine pores present, the resistance to attack increasing as this proportion decreases. Experience has shown that frost damage in the United Kingdom is unlikely in walls between eaves and damp-proof course (DPC) level. Whether this will continue to be true for highly insulated buildings, in which the outer leaf may be colder than hitherto, remains to be seen. Below DPC level and in parapets and free-standing walls, frost damage does occur and care is needed to select bricks of appropriate frost resistance. Information on such matters is available, in particular, from the Brick Development Association.

Many burnt clay bricks contain small amounts of salts, usually sulphates, and these may crystallise at the surface to give a white deposit referred to as efflorescence (Figure 6.1). This is noticeable, particularly, in dry periods following building and is usually harmless though, in exceptional cases, some

Figure 6.1 Efflorescence on brickwork.

crumbling of the surface may occur. This possibility is virtually restricted to underfired bricks, which not only present a weaker surface to attack but are also likely to contain larger amounts of deleterious salts. Some bricks may contain iron salts, which can produce rust stains and these are accentuated when such bricks are left exposed to rain before building. The stains affect appearance only and have no harmful effects on general durability.

Calcium silicate bricks are made not by burning clay but by reacting lime under steam pressure with a suitable source of silica, usually sand or crushed flint. In BS 187, calcium silicate bricks are classified partly by their drying shrinkage, which typically lies between 0.025 and 0.035 per cent. The coefficient of thermal expansion is around $14 \times 10^{-6}/°C$. Their movement is greater than of clay bricks, which necessitates a greater allowance for movement in the design of brickwork. Unlike burnt clay bricks, there is a marked correlation between frost resistance and strength. The stronger classes shown in BS 187 are required for the more exposed situations. Calcium silicate bricks are free from the salts which cause efflorescence in clay bricks. Exposure to atmospheric carbon dioxide causes slight hardening and shrinkage but is of little practical significance. Sulphur dioxide causes decomposition of the calcium silicate bonding material and this can cause general weakening and blistering in a very polluted environment. Sea spray is detrimental, causing erosion of the surface and also reducing resistance to frost attack.

Concrete bricks conforming to BS 6073 (Part 1) have drying shrinkage which may range from 0.04 to 0.06 per cent, with wetting expansion not greater than 0.06 to 0.08%. Roofing tiles are severely exposed. Their resistance to frost damage is dependent not only upon their intrinsic properties but also very much upon roof design. BS 473 includes a permeability test which provides a measure of frost resistance for concrete tiles. A water absorption test is an implicit measure in BS 402, which deals with clay roofing tiles. This Standard also calls for the rejection of tiles in which particles of lime are visible for, when wetted, these can cause local disruption of the tile.

6.1.5 Cement and concrete

Ordinary Portland cement is a material of major importance in construction. It is formed by burning ground limestone with clay and forms the basis of most concretes. As is well known, the cement sets and develops strength when mixed with water. The reactions which lead to this setting and strengthening are complex and are affected particularly by differences in manufacturing procedures and when admixtures are subsequently used. The chemistry of cement and concrete is treated comprehensively in specialist books and is not dealt with here. Other special cements are also made with specific properties, which are indicated broadly by their names, for example sulphate-resisting, low-heat, rapid-hardening, white, coloured, hydrophobic. Cements are also made from blast-furnace slag produced during iron manufacture and by the addition of pozzolanic material (usually pulverised fuel-ash) to ordinary Portland cement. When limestone and bauxite are used during manufacture instead of limestone and clay, high-alumina cement is produced, which is characterised by a rapid rate of strength development and, if not 'converted', has a marked resistance to attack by sulphates. Cement is not used on its own in construction. It is the durability and properties of the concrete, mortar or rendering in which the cement is used which are of main concern and the effects of external agencies on the durability of concrete are now considered.

Concrete which becomes wet and then dries alternately expands and contracts. The movement associated with wetting and drying is small, of the order of 0.04 per cent for normal dense concrete; it is attributed to the cement matrix. However, some aggregates in the United Kingdom and, particularly, in Scotland can themselves expand and contract on wetting and drying. Principally, these are igneous rocks of the basalt and dolerite types, and some sedimentary mudstones and greywackes. Moisture movement of concrete made with such aggregates may be at least doubled – and more than doubled if calcium chloride is used as an admixture. Movement around

0.08–0.1 per cent may not seem large, but it can lead to cracking of precast reinforced concrete units and to greater deflections of simply supported slabs. The use of such aggregates can also lead to a greater loss of pre-stress in pre-stressed concrete. It should be noted that the problem is mainly a Scottish one and most natural aggregates used in the United Kingdom do not show moisture movement of any significance. Design guidance on the shrinkage likely with these shrinkable aggregates, and the corresponding restrictions on usage, are given in Table 1 of BRE Digest 35.

Some forms of silica found in aggregates can, in the presence of water, react with alkalis derived from cement and, in so doing, may cause expansion and subsequent damage to the concrete. The reaction is known as alkali/aggregate reaction. The problem, well known in the United States, seemed to be absent in the United Kingdom, but a few cases have been reported in recent years from Jersey and the West Country, affecting a dam in the former case and the concrete bases of electricity substations in the latter. Since 1976, further trouble has occurred in the South-West and also in the Midlands, though it should be noted that the total amount of concrete affected is small. Several independent factors need to be present simultaneously before alkali/aggregate reaction can occur – the presence of reactive silica in the aggregate, sufficient sodium and potassium hydroxide released during the hydration of the cement, and the continued presence of water. The reactive silica leads to the formation of a gel which absorbs water to give a volume expansion which can cause disruption of the concrete. The risk of damage is slight with most aggregates and cement used, but reactive minerals have been found, particularly, though not exclusively, in sands and gravels dredged off the South Coast, the Bristol Channel and the Thames estuary.

The first signs of trouble take the form of fine random cracking and, when the alkali/silica reaction is severe, the presence of gel bordering the cracks which can be visible to the naked eye. The development of the reaction is slow. If the concrete will not be exposed to external moisture the likelihood of attack is negligible. In other cases, much will depend upon the alkali content of the cement and upon the aggregate used and reference should be made to the specialist literature (e.g. BRE Digest 258). Unfortunately, tests of a specialist nature outside the scope of this book are needed to detect whether or not an aggregate is likely to be reactive. A similar reaction can occur with some argillaceous dolomitic limestones known as alkali/carbonate reaction but no cases are known as yet in the United Kingdom.

Water freezing within the pores of concrete can cause disruption. Susceptibility to such attack is greatest with poor-quality concrete used in wholly exposed positions, such as kerbs and bridges. Good quality, low-permeability concrete, used in most building situations, is not affected. Frost resistance can be increased greatly through the judicious use of air-entraining agents. Water-soluble sulphates occur in some soils, notably the London, Oxford and Kimmeridge Clays, the Lower Lias and Keuper Marl. They may be present, too, in materials used as hardcore, for example plastered brick rubble. These sulphates of calcium, magnesium and sodium can attack the cement matrix to give reaction products which have an increased volume, and thus cause expansion. This, in turn, can lead not only to spalling and surface scaling but also to more serious disintegration. The extent of damage will depend greatly upon the amount and types of sulphate present, the groundwater conditions and the quality of the concrete. Once again, poor quality concrete will be affected more drastically than well-compacted concrete of low permeability. Considerable resistance to sulphate attack can be obtained by using sulphate-resisting Portland cement to BS 4027.

A form of sulphate attack identified in 1990 is thaumasite. Although the number of buildings and structures at risk of thaumasite sulphate attack (TSA) is small, building investigators should be on the lookout for this form of concrete decay. According to the Thaumasite Expert Group (1999),

> TSA will only occur in buried concretes when all the primary risk factors are present simultaneously and developed to a significant degree. It is emphasised that the probability of this occurring is concluded to be low for existing buildings and structures. These primary risk factors are:
>
> - Presence of sulphates/sulphides in the ground;
> - Presence of mobile groundwater;
> - Presence of carbonate, generally in the concrete aggregates;
> - Low temperatures (generally below 15°C).
>
> Additionally there are a number of secondary factors which influence the occurrence and severity of TSA and its effects:
>
> - Type and quality of cement used in concrete;
> - Quality of concrete;
> - Changes to ground chemistry and water regime resulting from construction;
> - Type, depth and geometry of buried concrete.

Steel reinforcement in concrete is inhibited from rusting by the high alkalinity of the surrounding concrete. Carbon dioxide, always present in the atmosphere, and sulphur dioxide reduce this alkalinity by carbonating the alkalis and thus increase the vulnerability of the steel to corrosion. In good-quality dense concrete, penetration by carbon dioxide is extremely slow and, when concrete cover is adequate, rusting may be almost indefinitely postponed. When concrete is permeable, or when cover is inadequate, carbonation can reach the depth of the reinforcement far more rapidly, with the consequent loss of rust inhibition. Corrosion of reinforcement or of pre-stressing steel can also occur if calcium chloride is used as an admixture, or if chloride occurs naturally in the aggregates used, and this can happen whether or not carbonation has taken place in the surrounding concrete. There have been many failures due to the use of calcium chloride, even when the total amount used was acceptable in the British Codes of Practice current at the time. Unfortunately, distribution within the mix can be uneven and harmful concentrations may occur. The problems of corrosion caused by the introduction of chlorides during mixing of the concrete are now more clearly recognised and restrictions are included in BS 8110.

Rusting of reinforcement and subsequent spalling, cracking and general disintegration of concrete are a hazard near coasts, where attack by sea spray can occur, and steel needs to have a good dense cover of concrete, at least 50 mm thick (Figure 6.2). Even unreinforced concrete can disintegrate if attacked heavily by sea salts. The corrosion of reinforcement will lead generally to cracking of the concrete in the direction of the reinforcement, the position of which can be determined by means of a cover meter. Rust stains near such cracks will also often be seen. Rust staining, however, may sometimes appear not through corrosion of the reinforcement but because of the presence of pyrite in the aggregate. The resistance of reinforced concrete to attack can be increased by using austenitic steels as the reinforcement (see Section 6.1.6), particularly where cover to the steel is lower than desirable as in, for example, cladding panels of thin section. Epoxy resin coatings to the reinforcement are also likely to improve resistance to corrosion, but only if control of the coating composition and its application are high. It is most important to avoid breaks in the coating system.

Concrete may also be made with lightweight aggregates, such as clinker, expanded clay or slate, foamed blast-furnace slag and sintered pulverised fuel ash. The drying shrinkage of structural concrete made with these aggregates is about double that of concrete made with gravel aggregate, and they are thus more liable to shrinkage cracking. Being more porous, lightweight concrete also offers less resistance to rusting of any steel reinforcement used. Some very lightweight concretes can be made using exfoliated vermiculite and expanded 'Perlite' as aggregates: drying shrinkage can be as high as 0.35 per cent. Such concretes are not used structurally, however, but for fire-resistance and thermal-insulation purposes.

High-alumina cement in concrete converts, as a result of a change in the structure of the hydrated cement, from an initial, meta-stable form to a more stable form which possesses, unfortunately, higher porosity and lower strength. The rate at which this change and weakening occurs depends upon both ambient temperature and humidity. At one time, it was believed that conversion was very slow at the temperatures and moisture conditions which could be expected in normal buildings. However, failures at a girls'

Figure 6.2 Corrosion of reinforcement causing cracking and spalling of concrete.

school in Camden in 1973 (DES 1973) and at a boys' school in Stepney the following year (Bate 1974) led to the conclusion that it would be advisable to assume that all high alumina concrete would reach a high level of conversion at some time in its life. Highly converted high-alumina concrete in the presence of water is vulnerable to chemical attack such as, for example, that which might be caused through contact with damp gypsum plaster. Under such circumstances, very low strengths are likely to be reached.

6.1.6 Metals

Aluminium, copper, lead, steel and zinc are the metals most commonly used in building. The choice of a metal, or an alloy, for a particular design situation, is dealt with extensively in the technical literature and in the large number of British Standards and Codes of Practice which govern their use. The most important agencies affecting performance are those which have a primary influence on corrosion – these are moisture, and gaseous, solid and liquid pollutants.

Under completely dry conditions, corrosion does not take place but, in most situations in buildings, moisture is present and corrosion is a potential risk. It is wise to assume that moisture will be present at some time or other. Moisture acts as an electrolyte and, often, a galvanic cell is formed which leads to the loss of metal forming the anode of that cell. The most common example of this situation arises when two different metals are in aqueous contact with one another. However, galvanic cells can be created when some metals are in moist contact with other building materials, such as bricks or plaster; when an alloy has two or more phases with different electrochemical characteristics (as may occur with some brasses); and even in single metals when a difference in oxygen concentration may occur at the surface, for example in pitted steel where the base of the pit has less access to oxygen than the metal surrounding the pit. Galvanic cells may also be created, and subsequent corrosion may occur, when particles of a metal or other substance are transported and deposited on other metals, as may happen in some plumbing systems. Corrosion generally is a complex electrochemical reaction which can be affected by the presence of dissolved atmospheric gaseous pollutants, by dirt, by manufactured admixtures (in particular, calcium chloride) and by temperature. Indeed, at some temperatures, bimetallic corrosion reactions may be reversed. The sensitivities to corrosion of the metals commonly used in building are described in the following.

(a) **Aluminium**

Aluminium is used mainly for cladding, flashings and window frames. Aluminium in contact with copper and its alloys is readily attacked, and severe damage can be caused. This can happen even at a distance, and examples are known of aluminium flashings and window frames suffering severe pitting corrosion through rainwater draining from copper-covered roofs and depositing small particles of copper on the aluminium. Similar attack can occur from lead. It is essential to use the correct aluminium alloy for a particular building purpose. Many prefabricated aluminium dwellings built in the 1950s used a high strength aluminium-copper-magnesium alloy. The presence of the copper, together with damp conditions caused by heavy and repeated condensation, led to serious corrosion (Figure 6.3). Wood preserved with copper-containing preservatives can also attack aluminium in contact with it and oak, too, can cause corrosion.

Figure 6.3 Corrosion of aluminium downpipe by copper salts.

Unprotected aluminium should not be embedded in cement mortars or in concrete, and direct exposure to sea spray will also lead to corrosion. Aluminium is not attacked by zinc, galvanised steel or stainless steel. Aluminium can be anodised by treating it electrochemically to thicken the natural oxide film which forms upon it during normal atmospheric exposure. This thickened film confers upon the metal an improved appearance and resistance to corrosion. The anodised layer is readily disfigured by splashes of cement or lime and needs to be well protected on site: many cases are known of badly marked aluminium cladding and window fitments from that cause.

(b) Copper

The main uses of copper are for roofing, cladding and plumbing. Copper is very resistant to corrosion and, when used in mixed metal systems likely to be encountered in building, it forms the non-corroding cathode. It can be attacked by flue gases containing sulphur dioxide if in close proximity to the chimney, but this can be readily prevented by good chimney design or by the use of a copper-silicon alloy. Copper is corroded by ammonia and by some mineral acids but the likelihood of such a combination of circumstances is small in building. However, it is well to keep copper from direct contact with latex cements used for fixing some types of flooring. Waters with a high carbon-dioxide content dissolves copper but the rate of loss is small and the effects on the copper are slight. Some years ago, copper tubing used for water supply suffered serious pitting corrosion. Carbon derived from the lubricant used in tube manufacture formed a film, which was nearly continuous and was cathodic to the copper surface. Pitting corrosion occurred at slight breaks in the film and caused rapid perforation of the tube. Nowadays, mechanical cleaning of tubes has largely eliminated this as a problem and tube conforming to BS 2871 should be satisfactory. The risk may still exist when imported tube has to be used in times of shortage of British tubing.

(c) Lead

Lead is used mainly for roofing (sometimes as a covering to steel sheets), for flashings and DPCs. It is highly resistant to corrosion through the formation of a dense, protective film of basic lead carbonate or sulphate. However, lead is not wholly immune from attack, and organic acids released from damp oak, Douglas fir and Western red cedar can cause corrosion: direct contact with these materials should be avoided. Lead is also attacked by free alkali present in rich cement/sand mortars, and severe corrosion has been known to occur within ten years with a DPC built-in with such a mortar. It is inadvisable to use unprotected lead sheet in such circumstances. Corrosion can be prevented by coating both sides with a bituminous paint.

A relatively new mode of roof covering failure is underside lead corrosion. This can be easily triggered by refurbishment works to a lead covered flat/pitched roof. In such circumstances, the opportunity to improve the roof's thermal insulation is usually taken. However, this can drastically increase the risk of underside corrosion to the lead. It is a problem usually caused by condensed moisture, which being relatively acidic can accelerate deterioration of this otherwise reasonably durable material. Organic acids from timbers, such as oak, near or in contact with the leadwork can exacerbate the corrosion.

As reported by Douglas (2006), the joint English Heritage/Lead Development Association's advice note for specifiers should be followed when undertaking major refurbishment work to a lead roof. In summary, the following general guidance is recommended by English Heritage:

- A ventilated warm roof design should be used. Adequately sized air spaces, ventilation inlets and outlets should be provided. This involves forming a 50 mm continuous vent gap *above* the warm deck insulation but below the lead sheeting and its deck. The 50 mm air space should provide full ventilation from eaves to ridge with no dead spots or obstructions (Figure 6.4).
- The existing roof structure must be dry before and during installation of lead coverings. For example, the timbers and other wooden parts of the roof system should have a moisture content below 18 per cent when the new covering and insulation is being installed.
- The underside of the lead should be coated with a protective passive film such as a slurry of chalk powder dispersed in distilled water.
- A sealed air and vapour barrier should be installed below the insulation.
- The substrate should be of low chemical reactivity. Any new timbers or decking material should be free from acidic preservatives and have a pH value above 5.5. Installing an inert fleecing layer or underlay between the existing timbers and the new covering can prevent these other aggressive chemicals from attacking the lead.
- Avoid using substrates comprising manufactured wood-based boards such as plywood, blockboard, chipboard, hardboard and orientated strand board, because of their chemical and moisture reactivity.
- The lead should ideally be installed in warm dry weather – say, May–July.
- Consider installing sensors (e.g. *H + R Curator*™ system) to monitor moisture and temperature conditions within vulnerable sections of the roof – for example, under outlets, gutters and so on.

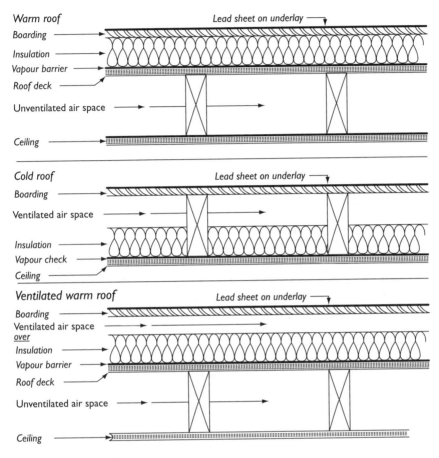

Figure 6.4 Various types of lead sheeted flat roofs.

(d) Steel

Steel is used in building for structural purposes, either on its own or as reinforcement in concrete; for window frames; for cladding and decking; for rainwater goods; and in plumbing systems. Ordinary mild steel, that is, steel with no substantial amounts of alloying metals, rusts when exposed to the atmosphere. Mild steel is seldom directly exposed but has a protective coating of paint, bitumen or zinc, or is enclosed by other materials, usually by concrete, which reduces the rate of rusting. The effects of the agencies of deterioration upon steel used as reinforcement in concrete have been considered already. Steel protected by zinc (i.e. galvanised steel) has a resistance to corrosion largely dependent upon that of the zinc. If steel is alloyed with small amounts of copper, about 0.25 per cent, the corrosion rate in air is roughly halved and the weathering appearance is improved, for the corrosion product takes the form of a more tenacious coating than ordinary flaky rust. A greater resistance to corrosion can be imparted when steel is alloyed with at least 10 per cent chromium, together with one or more other alloying elements. Steels so alloyed are known as stainless steels and the two main types used in building are ferritic and austenitic.

Ferritic stainless steels contain no appreciable amounts of nickel and some in the range have only modest resistance to corrosion. A ferritic stainless steel with 17 per cent chromium is used in the building industry but experience of its behaviour is, at present, limited. Austenitic stainless steels contain nickel as well as chromium and the two most commonly used in building have a nominal composition of 18–8 (18% chromium, 8% nickel) and 18–10–3 (18% chromium, 10% nickel and 3% molybdenum). The former has only modest resistance to corrosion and is suitable, principally, for internal use: exposed externally to moist, polluted air, pitting corrosion is likely. For cladding purposes, where good appearance is required, the 18–10–3 type is needed. Austenitic stainless steels are not affected by contact with most building materials and do not suffer in mixed metal systems, though they will act as the cathode in a mild steel/austenitic steel system and accelerate corrosion of the mild steel. An austenitic stainless steel is now used in domestic plumbing and heating. One problem that has arisen is corrosion under stress. Failures have occurred with stainless-steel back boilers in hard-water areas. Though the full sequence of the reaction is unclear, it is known that chlorides contained in many water supplies can concentrate behind the hard-water scale deposited in back boilers or in the air space when a boiler is not completely filled. At temperatures of around 80 °C, cracking known as stress corrosion cracking, may then occur – with, of course, failure of the boiler.

(e) **Zinc**

Zinc is used in building for roofing, cladding, flashings and, sometimes, for rainwater goods. Additionally, much of the total zinc usage is in the form of protective coatings to steel. These may be applied by several techniques, of which hot-dipped galvanising is the most common. Both hot dipping and metal spraying can achieve thick coatings at reasonable cost. The corrosion rate of zinc in unpolluted dry conditions is very slow and is of no practical significance in building. In unpolluted damp conditions, the rate may be four times as great and nearly ten times as great in damp polluted environments such as when zinc is exposed to sea spray or to sulphur gases or compounds. Zinc can be attacked by some wood preservatives and by inorganic flame-retardants. Rapid corrosion of galvanised wall ties has occurred when these have been embedded in black ash mortar, a source of sulphur compounds. Zinc is slightly attacked by wet concrete, but progressive attack is inhibited by the formation of calcium zincate, which is insoluble in the prevailing alkaline conditions. This resistance to attack has enabled galvanised steel to be used as reinforcement in concrete when enhanced protection is required, for example when difficulties are expected in providing the full depth of concrete cover normally expected. It should be noted, however, that bond strength can be reduced by hydrogen bubbles liberated in the initial reaction between zinc and alkalis in the concrete. To prevent this reduction, it is desirable to use galvanised reinforcement which has been 'passivated' by a chromate treatment. Galvanised steel is also used for cold-water cisterns. Soft water, which contains dissolved oxygen and carbon dioxide, attacks zinc which, under such circumstances, will need protection by, for example, two coats of bituminous paint. Severe pitting corrosion will occurr when copper and brass debris from plumbing installation activities have been left inside unprotected galvanised cisterns and this should always be prevented. Zinc alloyed with copper can provide a range of brasses and one such, known as alpha-beta brass, is commonly used for fittings in water services. If the water supply is acidic, or is an alkaline water with a high chloride content, zinc can be removed from the brass (dezincification), leaving behind a spongy copper residue. This can lead either to penetration of the wall of the tap or fitting, allowing water seepage, or to their blockage by the zinc corrosion products. If such water supplies cannot be treated to increase their temporary hardness, which will prevent the attack, the use of hot-pressed alpha-beta brass fittings should be avoided. Zinc cladding, roofing and flashings can be corroded by oak, Douglas fir and Western red cedar, and so direct contact should be avoided.

(f) **Fatigue and Creep**

While corrosion remains the principal cause of deterioration of metals, some problems of fatigue and creep are worth noting. Metals subjected to a steady load deform slowly: movement depends upon load, temperature and time. This is known as creep and is not, in general, serious. Failures of lead flashings have occurred, however, partly as a result of the low resistance of lead to creep but more through its high thermal movement and low resistance to fatigue. This tendency can be inhibited by good design, with appropriate limitations on the size of any one piece of sheet lead used, or by the use of lead containing a small proportion of copper. Technical recommendations are incorporated by the Lead Development Association in their technical guides. A zinc/titanium alloy is also available which has a greater resistance to creep than unalloyed zinc.

6.1.7 Glass

(a) Ordinary glass failure

Ordinary float or plate glass is a very durable material and is seldom affected by any of the agencies of deterioration mentioned previously. Surface etching can occur if sheets are closely stacked under damp conditions, and if alkali from paint removers splashes on to glass and is not removed. Thermal stresses can cause cracking, chiefly because glass has a rather different coefficient of thermal expansion than common framing materials. Proper allowance must be made in design and construction to accommodate differential movement and this aspect is dealt with in more detail in the section on cladding.

(b) Toughened glass failure

Although toughened glass is a material with strength and stiffness similar to aluminium, ironically, it is not as problem-free as one might think. Nickel sulphide–induced failure is a major modern mode of failure of toughened glass. It's a spontaneous failure caused by nickel sulphide inclusions in the glass and are almost impossible to spot with the naked eye.

After manufacture, toughened glass is frequently heat soaked. However, this process does not always prevent glass panels with nickel sulphide crystals being installed in the building. The nickel sulphide crystals can grow in size with time and cause the glass to shatter into small fragments without warning.

Wheeler (2002) reported on this phenomenon after a spate of toughened glass failures involving nickel sulphide inclusion:

> Toughened glass, sometimes referred to as tempered glass, is produced by heating ordinary annealed glass to just below its softening point, about 650 °C, and then cooling it rapidly with blasts of air. This causes the surface of the glass to cool more rapidly than the inner core, which in turn causes the outer zones of the glass to be in compression while the inner zone is in tension. The compressive and tensile stresses balance one another and in doing so impart to the glass a bend strength some four to five times greater than ordinary glass. The increased bend strength provides resistance to breakage. In addition, when glass toughened in accordance with the provisions of BS6206 breaks, it does so "safely". In effect, the glass dices into small pieces with a diameter no larger than 10 mm and without sharp edges. The standards and codes of practice, however, deal with safety in relation to human impact; they do not deal with the effect that broken safety glass may have when it falls from height.

Two relevant principal causes for glass failure have been identified. First, "hard point glass to metal contact" was identified, where glass was directly bearing onto the glazing bars. This was found to be occurring in several places. Second, "inclusion expansions" were noted. In cases where contamination of raw materials (or melt) by nickel containing metallic alloys, eg nichrome wire or thermocouples, occurs, the nickel and sulphur can react under the high furnace temperatures and produce nickel sulphide inclusions.

As toughened glass has outer surfaces that are in compression and an inner core that is in tension, the compressive and tensile forces are in equilibrium. If, however, an unstable inclusion particle is located within the tension zone of the glass the tension will be increased as the changes occur. If the size of the unstable inclusion is appropriate, the increase in tension will reach a critical level. A micro-crack will develop at the inclusion site (fracture origin) causing imbalance and followed by spontaneous failure of the glass. Three conditions are needed for this type of failure:

- an unstable nickel sulphide inclusion;
- a particle size greater than 80μ; and
- a location within the tension zone of the glass.

Failure of toughened glass usually causes great inconvenience to building owners and sometimes may also pose a serious risk of injury to the occupants and passers by. However, the dominant form of toughened glass failure is not spontaneous failure. The other possible causes of tempered glass failure are

- sudden impact;
- overstressing caused by poor handling or improper fixing; and
- heavy usage resulting in glass pane slowly slipping-off from the door hinges and falling.

6.1.8 Mortar and renderings

Most present-day mortars consist of cement and sand – with or without an air-entraining agent – or cement, lime and sand. Mix proportions depend upon the types of brick and block used and their exposure. The effects of shrinkage caused by drying and by carbonation are generally insignificant. Mortars can be affected by frost but their resistance can be increased through the use of an air-entraining agent.

Serious disintegration of mortar can occur when soluble sulphates, sometimes present in wet brickwork, react with tricalcium aluminate, present in ordinary Portland cement mortars. A considerable increase in the volume of the mortar can then occur which causes the mortar to split and to become friable (see Chapter 10). Similar sulphoaluminate attack can occur when mortar is exposed to condensed water vapour containing sulphates derived from flue gases. This was a particular hazard when slow-combustion fuel appliances were used with unlined chimneys. Typical external renderings are broadly similar in composition to mortars and have similar properties. Sulphate attack can occur through salts derived from wet brickwork: frost attack is rare. Drying shrinkage can be more of a problem and strong cement/sand mixes can craze and crack, particularly if applied in warm, dry weather.

6.1.9 Plastics

A wide range of plastics is used in building. PVC has the widest application, in either a plasticised or unplasticised form. Plasticised PVC is extensively used as a floor covering, as a sarking under pitched roofs, as a membrane for covering flat roofs and in the manufacture of water stops in concrete structures. Rigid unplasticised PVC is used principally for domestic soil and vent systems, for rainwater disposal and drainage, for wall cladding, as translucent or opaque corrugated roof sheeting and for ducting, skirtings and architraves. Rigid PVC is also used for window frames, sometimes in combination with metal or timber. Expanded PVC is available in board form for thermal insulation.

Polyester resins, reinforced with glass fibre, are used principally not only as cladding sheets but also for cold-water cisterns, in water and sewage disposal systems and for industrial gutters. Polyethylene, too, is used for cold-water cisterns and floats, for domestic cold-water pipes and for bath, basin and sink waste pipes. In sheet form, it is used as a damp-proof membrane and for covering concrete and hardcore surfaces.

Acrylic resins are used for sinks, drainers and baths, for corrugated sheeting and for roof lights. Other plastics employed significantly in the building industry include

- crylonitrile butadiene styrene copolymers for large drainage chambers;
- polypropylene for plumbing and drainage fittings and as wall ties;
- polybutylene for hot and cold water supply pipes;
- polycarbonates for glazing;
- phenol formaldehyde resins used to impregnate paper and fabric to provide wall and roof sheets, and foamed to give a cellular material used for thermal insulation;
- epoxide resins for in situ flooring and for concrete repair; and
- and polystyrene, polyurethane and urea-formaldehyde in expanded form for thermal insulation.

This is not a complete list, for many other plastics can find some use in the building industry, for example nylon, cellulose acetate and polyacetals are used in taps and miscellaneous fittings. Short-wave solar radiation degrades plastics by causing embrittlement and a change in surface appearance. The risk is greatest in coastal and rural areas. The effects can be reduced or increased by additives incorporated into the plastic. The addition of fire retardants reduces resistance to degradation. On the evidence so far available, the durability of plastics seems to be good. Moisture in general has little effect but can reduce bond strength between glass fibre and polyester resin: the extent of any weakening depends greatly upon the control exercised in manufacture. Plastics are not harmed, in general, by contact with other building materials, though cracking of polyethylene cold-water cisterns has been caused by the use of oil-based jointing compounds.

As will be seen from Table 6.2, PVC and polycarbonates have high thermal expansion. Unless properly allowed for, the movement of PVC gutters and down pipes can cause joint failure and leakage. Varieties of polyethylene have even greater thermal movement. Plastics creep under continued loads and special precautions are needed when stresses are high, as in filled cold-water cisterns – for example, by complete support of the base of the cistern. The long-term behaviour of many of the plastics used is to some extent uncertain, for related experience is not yet available. More work is needed to enable better prediction of fatigue and creep under long-term loading. However, external performance so far has been generally good and for plastics used internally, or otherwise shielded from direct sunlight, the effects of short-wave solar radiation are nullified.

6.1.10 Natural stone

Natural stones are classified as belonging to one of three main groups – igneous, sedimentary or metamorphic. Igneous rocks are formed by the solidification of a molten magma: common examples are granite, dolerite, basalt and pumice. The only one used to any extent in massive form in building is granite, though many may be used in crushed form as aggregates. Granite is highly resistant to all the agencies of deterioration described in Chapter 5; thermal movement is unexceptional and there is no moisture movement. Sedimentary rocks may derive from particles produced from older rocks by the normal processes of weathering, from the accumulation of skeletons (usually marine organisms) and by chemical deposition. The particles forming the rock are deposited as sediments, through the action of water and wind, and are cemented together to varying degrees by minerals. Consolidation is assisted by pressure arising from the mass of the sediments as they build up in thickness. Sedimentary rocks have a natural bed though this may not always be apparent. Limestone and sandstone are the principal examples of sedimentary rocks and provide most of the building stone used in the United Kingdom. Most limestones consist essentially of calcium carbonate but a proportion of magnesium carbonate may also be present and, when this is significant, the limestones may be called magnesian limestones. Sandstones consist essentially of grains of quartz cemented together by silica (siliceous sandstones), calcium carbonate (calcareous sandstones) or, sometimes, by both calcium carbonate and magnesium carbonate (dolomitic sandstones). Sandstones containing appreciable quantities of oxides of iron are termed ferruginous.

The main cause of the deterioration of limestones and sandstones is atmospheric pollution. Sulphur gases dissolved in rainwater react with calcium carbonate to form calcium sulphate. When this crystallises under dry conditions, it generates a stress which can break off particles of stone of varying size. If wetting is frequent from rainfall, the surface of the stone is slowly eroded and the calcium sulphate is continually removed. Under sheltered conditions, the calcium sulphate, which is only sparingly soluble, may build up to form a hard skin which causes more unsightly damage. When the skin eventually blisters and breaks off, it pulls away limestone with it. Dissolved sulphur gases can also remove the bonding calcium carbonate in calcareous sandstones and, in so doing, severely weaken the stone. Magnesium carbonate is also attacked by sulphur gases leading to similar deterioration of magnesian limestones and sandstones. Ground salts, and those in sea spray, can also cause damage. The former can cause expansive damage when penetrating into limestones and calcareous sandstones. Attack from sea spray is usually manifested by a general powdering of the external surfaces. Frost may attack some limestones, though rarely between eaves and DPC level: British sandstones are virtually immune to attack. The resistance of both limestones and sandstones to attack by frost, and to the crystallisation of salts, depends essentially upon the pore structure of the stone. Resistance depends, particularly, upon the proportion of fine pores present and, to a lesser extent, upon total porosity. As the proportion of fine pores present increases, resistance to attack decreases, but this is a guide rather than a rule, and experience of use or specialist advice are the best aids to specification. A recent publication suggests zones within a building, for example steps, copings, plinths and plain walling, within which the more common British limestones can safely be used given a particular environment (BRE Digest 269).

A phenomenon known as contour scaling can cause damage to sandstones (Figure 6.5). Calcium sulphate can be deposited within the pores of sandstone, even though the latter may not contain calcium carbonate. The calcium sulphate may derive from limestones attacked by sulphur gases which has then washed on to the sandstone surface, or has been formed from some other source of calcium. Whatever its derivation, internal stresses are created in the surface crust and these are thought to arise from differences in

Figure 6.5 Contour scaling of sandstone.

thermal expansion and/or moisture movement between the sandstone and the crust blocked with calcium sulphate. The crust breaks away, usually at a depth of between 5 and 20 mm, and follows the contours of the surface (BRE Digest 177).

Any deterioration through pollution, salts and frost is likely to be enhanced if sedimentary stones are laid so that the natural bed lies parallel to the vertical face of the wall. This is known as face bedding and should be avoided. The greatest durability is achieved when the natural bedding plane lies parallel to the horizontal courses in a wall. Metamorphic rocks are derived from sedimentary rocks through the action upon them by great heat and pressure arising from movements of the Earth's crust. The structure of the rock is then changed radically. The only metamorphic rocks of significance for building purposes in the United Kingdom are marble (derived from limestone) and slate (derived from clays). Marble is used for cladding and is very durable: it is not immune from attack by sulphur gases but this is rare in reality. Slates are used for roofing, for cladding and as DPCs. Roofing slates are exposed to the most severe conditions and can be affected by sulphur gases. Those slates conforming to BS 680, however, are highly resistant to attack. The Standard includes a sulphuric acid immersion test which is intended to simulate, but accelerate, the attack by sulphur gases dissolved in rainwater on any calcium carbonate present in slates. One major cause of damage to all stones can arise if the fixings embedded in them corrode. In the past, the rusting of iron and steel cramps and dowels has caused extensive damage, particularly to limestones and sandstones. Appropriate metals to use are stated in BS 5390.

6.1.11 Timber

There are many species of wood available that can be used for building purposes. Extensive technical data on the types, uses and properties of different woods are published in standard reference books and, notably, in the publications of the Princes Risborough Laboratory and the Timber Research and Development Association. Softwoods are used for most structural and joinery work, and hardwoods for more specialised purposes, such as decorative flooring. The terms "softwood" and "hardwood" are not of particular significance in building and should not be taken as relating directly to strength and hardness. Most of the softwoods used belong to the species *Pinus sylvestris* (Scots pine, Baltic redwood), *Picea abies* (Norway spruce, European spruce, Baltic whitewood) and *Pseudotsuga menziesii* (Douglas fir, Columbian pine, Oregon pine). Larch, Western hemlock and Western red cedar are notable amongst other softwoods used. The agencies which bear most upon the performance of timber are moulds, fungi, insects and moisture. The principal effects of these on timber are dealt with more conveniently in Chapter 5 and require little further comment. While sapwood is more susceptible to attack than heartwood it should be noted that even the heartwood of redwood, whitewood and Western hemlock is not naturally durable. Moreover, market economics result in the presence of sapwood in most softwood available. Suitable preservative treatments are needed where there is any risk of attack by the agencies mentioned. When seasoned timber absorbs and loses moisture it expands and contracts, and these moisture movements are much higher than for any other building material. Moisture movement in timber is a complex phenomenon and is affected by many factors, which include the speed of drying and wetting, the temperature at which the timber is seasoned, the direction in which the movement occurs and, of course, the species of timber. Tangential movement, i.e. movement parallel to the annual rings, is greatest, followed by radial movement, i.e. movement at right angles to the annual rings. Longitudinal movement in the length of the timber is smallest and, generally, disregarded. A standard method of measurement of moisture movement is to record the dimensional change, which occurs when timber conditioned to equilibrium at 90 per cent rh (relative humidity) and 25 °C is dried to equilibrium at 60 per cent rh and 25 °C (Princes Risborough Research Laboratory 1982). The movement is expressed as a percentage of the width of the timber specimen in this latter condition. Some typical tangential and radial movements for timbers commonly used in building are given in Table 6.1.

Timber subjected to changes in moisture can deform by bowing, twisting or cupping, and may crack if movements are frequent. Total moisture movement is not a sound guide, however, to the likelihood of distortion, for other factors can influence this, for example grain direction and the presence of knots. To reduce the risk of distortion, good practice requires that timber is seasoned to around the average moisture conditions likely to be met in its place of use. BS 5268 Part 2 relates these moisture contents to positions within a structure. Timber is an organic material and, as such, is affected by prolonged exposure to sunlight. Short-wave solar radiation causes degradation of the surface, and timber stored on building sites unprotected from rain and sun acquires a typical grey appearance. Some fungi can discolour wood without affecting its general durability and strength. One such fungus is called blue-stain, which can stain undesirably the sapwood of some softwood. This is a problem encountered before use, however, for the fungus does not grow on timber used at moisture levels considered safe in buildings. If excessive moisture remains in any timber used, attack by far more serious fungi is likely to develop, as already described.

6.1.12 Moisture movement

Many of the building materials mentioned in the preceding paragraphs expand and contract when they take up or lose moisture. This movement is not always wholly reversible but usually approximately so. It depends, inter alia, upon the degree of wetting and drying, the precise composition of the material, and the plane of measurement. Typical movements shown in Table 6.1 are those which could be caused by changes in moisture content likely to be encountered in environments common to normal buildings.

This should be taken as a guide only and the specialist literature should be consulted if more precise values of movement under defined changes in the moisture regime are sought.

6.2   Service life assessment

6.2.1   Background

Predicting the service life of a component, material or product is an art as well as a science. It is not an exact discipline, however, owing to the large number of variables and uncertainties involved.

It is important to recognise that there are various types of lives of a building. These are summarised as follows

- *Aesthetic life*: The time within which the component retains its initial (or near) original appearance.
- *Design life*: It is the period of use as intended by the designer – for example, as stated by the designer to the client to support specification decisions (BS 7543).
- *Economic life*: The time within which the asset is worth more than its site value or opportunity cost.
- *Physical life*: The time it takes for a building, subsystem or component to wear out.

6.2.2   Service life determinants

The principal service life determinants are outlined in Table 6.2.

6.2.3   Sources of service life data

The three main sources of service life data are

- Experience: Past projects involving identical materials under similar conditions. Texts such as the HAPM Life Component Guide (2000) or the NBA guide (1985) offer useful sources of information on estimated life expectancy of common materials, components and elements.
- Modelling: Mathematical analyses, usually involving statistical assessments of predicted failure patterns.
- Testing: Field test involving actual exposure conditions or laboratory work involving accelerated tests.

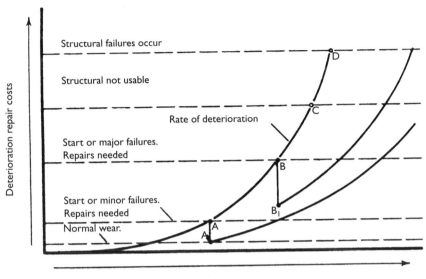

Figure 6.6 A curve showing deterioration rate and time relationship for maintained and unmaintained buildings. Points A through D represent stages of accelerated deterioration. Structures repaired at point A cost less overall and last longer than structures repaired at point B (compare curve A_1 to B_1). *(Courtesy: Walker Parking Consultants/Engineers, Inc.)*

6.2.4 Anticipated life expectancy of components

Given the points made in the previous section, a major premise in Building Pathology is that nothing physical lasts forever. With this in mind, therefore, it is likely that all elements of a building will require some form of maintenance at some time during their service life. The timescale involved in such aftercare may vary from daily servicing to a major replacement after many years (Figure 6.6).

Sources of typical decoration and renewal cycles are

- NBA (1983)
- Scottish Office/Scottish Homes (1989).

6.2.5 Methodology of service life prediction

The procedure for estimating the service life of a component or element is based on the following factors

- *Performance characteristics*: thermal/acoustic insulation, fire resistance, minimum life expectancy, and so on.

- *Exposure conditions*: Sheltered/exposed, coastal/inland.
- *Site examination*: Visual inspection, non-destructive examination, extraction of samples for laboratory tests.
- *Recall important dates*: Maintenance record, adaptations.
- *Undertake or analyse tests*: Field tests or laboratory tests.
- *Data analysis*: Moisture readings, crack sizes, changes in form.
- *Result*: Premature degradation, remaining expected service life.

See ISO15686: 2000 (Buildings and constructed assets – Service life planning. Part A: General principles; Part B: Service life prediction procedures; Part C: Performance audits and reviews) on the background to and methodology of service life prediction.

6.2.6   Whole life costing

A technique that compliments service life assessment is whole life costing (see Boussabaine 2004). Both of them can be used in the prognosis rather than diagnosis of building defects. According to the BRE (2002a), service life assessment and whole life costing have characteristics shown in Figure 6.7.

See OGC (2003) guide to the use of whole life cycle costing in construction.

6.2.7   Deterioration rates

Predicting the rate of deterioration of buildings and their elements is not an exact science. It is influenced by several variables, such as environmental conditions, quality of material, standard of workmanship, and so on.

For simplicity, it can be assumed initially that the rate of deterioration of a component or element is linear. Adaptation (e.g. refurbishment) is usually the

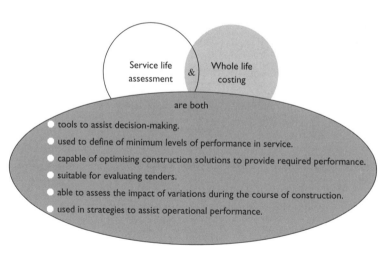

Figure 6.7 Service life assessment and whole life costing characteristics (BRE 2002a).

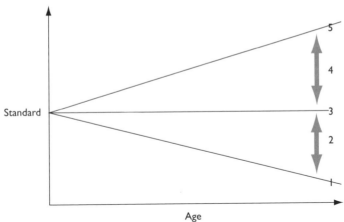

1. Deterioration
2. Works carried out to restore to original or near original standard = maintenance
3. Original standard
4. Increasing performance required by occupiers
5. Works carried out to achieve higher standard than original = adaptation

Figure 6.8 The relationship between deterioration and user expectations.

necessary action to offset this effect (Figure 6.8). However, there are different forms of ageing other than linear: abrupt and exponential – depending on the exposure and use circumstances (Douglas 2006) (Figure 6.9).

6.2.8 Failure rates of services

The bathtub curve can be used as a means of describing the likelihood of the instantaneous failure of a services unit (boiler, pump, etc.) or system (air conditioning, heating, etc.) over its whole life. It derives its name from its typical shape, which resembles a bathtub (Figure 6.10). It plots the likelihood of failure (failure rate) against time, and is the overall curve that combines early failures, random failures and wearout failures.

There are three phases in the bathtub failure curve, each corresponding to the burn-in, useful life and burn-out sections:

- Phase 1: Burn-in, or early (i.e. decreasing) failure learning period – the downward sloping part of the curve.
- Phase 2: Useful life, or constant failure rate – the flat part of the curve. Its length of period depends on frequency and quality of maintenance.
- Phase 3: Burn-out/wearout, or increasing failure rate of long-life items – the upward sloping part of the curve. Wearing-out period. Tolerances absorbed.

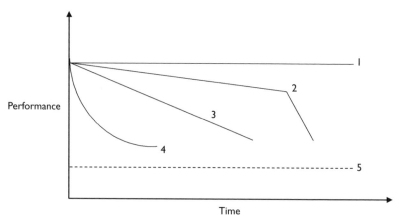

Notes
1. Ideal performance profile
2. Abrupt
3. Linear
4. Exponential
5. Minimum acceptable standard

Figure 6.9 Different types of ageing.

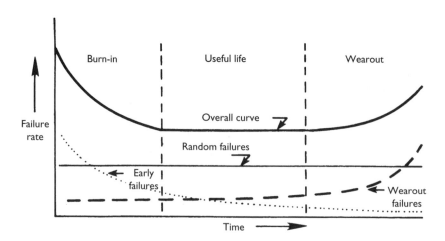

Figure 6.10 Bathtub failure rate curve.

6.3   Summary

The ability to resist wear and tear and other degradation mechanisms is an important performance characteristic of all building products. Defects will compromise the durability of building elements and thus shorten their service life.

Chapter 7

Moisture

OVERVIEW

This chapter addresses the problems caused by moisture in buildings. It examines the key ways in which this, the most dangerous of all catalysts in buildings, can cause trouble.

7.1 Introduction

The special position of moisture as the principal agent causing deterioration was mentioned in Chapter 6. Moisture, of course, cannot be totally excluded from buildings. It's needed for both the building itself and for the comfort of its occupants. Organic materials such as timber devoid of moisture would be affected by desiccation. This in turn would cause them to shrink, crack and warp. The occupants, too, cannot inhabit a totally dry building (i.e. one in which the relatively humidity is near or at 0%). Excessively dry conditions would make the occupants' nose, eyes and mouth parched and thus uncomfortable, increasing their risk of coughing and suffering from ear, nose and throat problems. The risk of electrostatic shocks increases dramatically if the relative humidity of the indoor air goes below 30 per cent (see BSI 1991b).

Too much moisture (i.e. >70% rh) in a building can be damaging not only to its structure, fabric and contents but also to its occupants. The risks to health associated with dampness are well established (Douglas 2006; Howieson 2005).

A knowledge of the main sources of moisture therefore is vital (see Trotman *et al.* 2004). It will enable steps to be planned to minimise the total amounts trapped, entering or generated in a building, and so help reduce or prevent some of its detrimental influences (Rose 2005).

7.2 Sources of moisture

7.2.1 Water entering during construction

Much water is used during construction for the mixing of mortar, concrete and plaster and for wetting bricks before laying. For brickwork alone, as much as one tonne of water may be used in building the average house (Addleson and Rice 1991). Some of this constructional water is immobilised in the hydration of cement and plaster, and some evaporates before occupation of the building. Even so, much water necessarily used for mixing purposes will still be retained and can be slow to dry. Not a lot can be done about this moisture, though good ventilation and the judicious use of heating in the first year will assist drying. Unfortunately, a good deal of water is introduced quite unnecessarily by poor site control of the storage of materials and by inadequate protection of partially built structures. It is all too common to see timber roofing and joinery, blocks intended for internal use, and stacks of bricks left wholly unprotected from the rain. These components are then installed wet and often left uncovered, so that further rain keeps them wet. Flat roof decks, especially those screeded with lightweight concrete, can absorb large quantities of rainwater if not adequately covered during the constructional phase. An average house, with masonry walls, is likely to contain several tonnes of water just after completion when site control is of a standard commonly seen today (Addleson and Rice 1992). At least a year is likely to elapse after occupation before the moisture level drops to that in equilibrium with normal internal humidity conditions. Efficient site storage and protection could go a good way towards reducing these amounts. In particular, the tops of walls, floors and roofs should be protected, as far as possible, and all timber completely protected from rain and ground water.

A surface damp proof membrane (such as *TREMCO's ES 2000*) is often used in new construction. It can prove very effective in dealing with the problems of laying a floor covering on a "green" concrete slab or screed (Pye and Harrison 1997).

7.2.2 Ground water

In the United Kingdom, the level of the water table is seldom far below the surface of the ground. Materials in contact with the ground will draw up this water by capillary action into their pores and into the structure of which they are a part. It should not be assumed that this will not happen, even on an apparently dry site. Building operations undertaken below ground can, in themselves, change, sometimes detrimentally, the pattern of natural water drainage and also the level of the water table. The height to which ground water can rise if not obstructed can be considerable: it is affected by a number of factors, chief of which are the pore structure of the materials in contact with the ground and the depth below ground of the water table. These ground waters contain salts in solution which tend to concentrate on wall surfaces. The salts are often hygroscopic and so add to general dampness by absorbing moisture from humid air. It might be thought that entry of ground water and its associated salts is a problem of the past – and so it should be, since the incorporation of a DPC was made compulsory in 1875. Unfortunately, cases of attack are not infrequent in new buildings and arise, generally, not because a DPC has been omitted but because it has been bridged and so made ineffective. Common causes of bridging are considered in the chapter dealing with floors and walls.

The passage of ground water into a structure should be suspected when internal wall surfaces show persistent dampness in an irregular pattern, to a height which can, on occasions, exceed 600 mm (Figure 7.1). The greatest heights reached occur on damp sites and when the outer faces of the external walls do not allow easy drying, for example if covered with a non-porous coating. When salts are present, decorations may be pushed off. The salts are usually left behind and seen as a white powdery growth (Burkinshaw and Parrett 2003).

Figure 7.1 Stains typical of damp penetration at ground floor level.

7.2.3 Rain and snow

From the earliest times, a primary purpose of any building has been to shelter the occupants from rain and snow, and it is disturbing that so many buildings fail to do so adequately. Rain and snow may penetrate directly through gaps in the structure, particularly at the junction of windows and doors with walls, at joints between cladding panels and at gaps in sarking felts. It may enter indirectly by passage through porous building materials under the action of capillary forces. Passage may be assisted by gravitational forces, as when water penetrates through a flat roof. Indirect penetration commonly occurs through blocked cavity walls, through solid walls ineffectively rendered or pointed, and particularly through flat concrete roofs. The risk of rain penetration through gaps in a structure is greater when the rain is wind-assisted than in still conditions. The severity of exposure to rain was seen as likely to vary both with rainfall and with wind speed: measurements made by Lacy (1976), using rain gauges set in the walls of buildings, showed that the amount of rain driven on to a wall was directly proportional to the product of the rainfall on the ground and the wind speed during the rain. This relationship was used as the basis for the derivation of an index of driving rain which could be used to give relative probabilities of rain penetration. The product of average annual total rainfall in metres and annual wind speed in metres per second provides the annual mean index of exposure to driving rain. Contour maps have been produced showing the annual mean driving-rain index for areas in the British Isles. The wind speed used in the calculation of the index refers to an open, level site, so local correction factors have been derived which allow for local topography, for the roughness of the terrain and for the height from the ground to the top of the building (Lacy 1976). These factors are designed to allow local driving-rain index values to be obtained which relate to the amount of rain penetration through walls that actually occurs.

Computer analysis of meteorological data has, recently, enabled the Meteorological Office to produce more realistic values based on the fact that heavy rain is usually associated with strong winds. Two indices have consequently been derived. The local annual index measures total rainfall in a year and is most significant for the average moisture content of masonry. The local spell index measures maximum intensity in a given period and is most significant for rain penetration through masonry (BSI 1984b).

The amount of rain blowing on to walls can be considerable: even in moderate conditions, up to one tonne per square metre a year can be so driven. The pattern of wetting of the walls of any individual building will vary with its fabric and the extent to which it controls absorption and run-off – and, of course, with individual design features (Figure 7.2). As a generality, it can be said that driving rain tends to concentrate at internal and external angles and, particularly, at the edges of buildings. The extent to which walls absorb rain depends much upon the pore structure of the

Figure 7.2 How not to dispose of rainwater.

materials used – there can be a difference of 100 to 1 between types of bricks. The rate of drying after wetting is greatly dependent upon pore structure and upon the ambient temperature, humidity and wind speed. The extent of drying depends on the frequency and duration of the dry spells between periods of rain. Joints between non-porous components are particularly vulnerable to penetration by rain or snow under wind pressure, for they are unable to hold moisture by absorption and so retard its passage to the interior. When rain penetrates a crack in a relatively impervious system, for example in a dense rendering, it can be slow to evaporate and can cause many of the problems associated with rain penetration which might have been avoided had free evaporation been possible.

Falling snow has a density typically only one-tenth that of rain. It can, as a consequence, be blown upwards to a greater extent and also distributed unevenly on structures. It is not uncommon for it to be blown under eaves and sarking and to build up so that, on melting, considerable wetting of roof timbers takes place. Although it is not a major problem in the United Kingdom, the effect of eccentricity of snow loads on pitched roofs needs to be consciously considered in design. Heavy snow accompanied by high winds and the subsequent drifting has caused partial or total collapse. Guidance on the high local loads that drifting snow can impose is given in BRE Digest 290. Direct penetration of rain or melted snow shows as damp patches, varying in size and position, on the inside face of external walls, usually within a few hours of the beginning of precipitation. These patches appear most often around window and external door frames, but not exclusively so.

Competent building design and construction should prevent such penetration. Many of the traditional ways of keeping rain away from the main structure, for example by the use of a good eaves overhang and by recessing windows, seem to have been abandoned in most modern dwellings, possibly through ignorance, indifference or false economy. They need to be reintroduced.

7.3   Moisture from human activities

A great deal of moisture is produced from normal human activities and this can be a major input to help cause condensation (Garratt and Nowak 1991), which is considered in some detail in the next section. Just by normal breathing, one person produces at least 0.3 l of moisture in a day. Typical domestic activities can greatly exceed this amount. Clothes washing and drying constitute a major source of moisture input. Drying a normal wash for a family of five can generate as much as 5 litres of moisture, some ten times as much as is likely from the washing phase. To add to the difficulties, the drying of clothes indoors is likely to occur most often when the weather is damp. It is not uncommon, too, for large volumes of water to be left in baths and sinks, for considerable periods of time, to soak crockery and clothes. More than half the daily input of moisture in a typical home is produced in the kitchen.

Unvented heaters, such as paraffin stoves and free-standing gas appliances, including gas ovens, which are often used as supplementary heaters, generate large amounts of moisture. Paraffin produces more moisture than fuel consumed – approximately 1.2 l per litre of paraffin. The use of such heaters has grown in recent years, as their fuel costs are usually lower than the cost of electricity. Table 7.1, based on data in BS 5250 (BSI 2002b), shows some

Table 7.1 Typical daily moisture production within a five-person dwelling. (Based on data contained in BS 5250: 2002b. The figures do not take into account any moisture removed by ventilation.)

Activity	Moisture emission (litres)
Clothes drying	5.0
Cooking	3.0
Paraffin heater	1.7
Two persons active for 16 hours	1.7
Five persons asleep for 8 hours	1.5
Bathing, dishwashing	1.0
Clothes washing	0.5
Total	14.4

moisture outputs from typical domestic situations and the pre-eminence of clothes drying as a producer of moisture.

7.4   Condensation

The presence of moisture may lead not only to direct attack on materials but also to condensation. Condensation occurs when warm moist air lands on a surface that has a temperature at or below dew point. This results in moisture in vapour form being converted to liquid. The beads of water on single glazing in kitchens and toilets in use are a typical symptom of surface condensation.

It is probable that condensation and its effects have been the greatest single post-war problem in building, particularly in local-authority housing. Condensation and mould growth now affect some two and a half million homes in the United Kingdom (Research, Analysis & Evaluation Division 2003; House Condition Surveys Team 2004).

The cause of condensation is relatively simple. Air at any time will contain some water vapour and warm air is able to hold more water vapour than cold air. When air containing moisture is cooled progressively, there will come a temperature at which the air cannot hold all the moisture any longer and it is then said to be saturated. This related temperature is known as the dew-point and is the temperature below which condensation will begin. Water vapour in air exerts a pressure, and this pressure causes it to move through all except completely impermeable materials to areas of lower vapour pressure. At any given temperature, the ratio of the actual vapour pressure exerted to the vapour pressure which would be exerted if the air were saturated is known as the relative humidity (rh); the higher the rh, the closer the air is to saturation.

The relationships between vapour pressure, dew-point, rh, temperature and the actual amounts of water contained in air can be represented by a psychrometric chart such as that shown in Figure 7.3. This looks complicated but two examples will suffice to show that its use is simple. If the air within a room at a typical temperature of, say, 20 °C has a relative humidity of 40 per cent (point A) and this air is then cooled, reading horizontally to the left, point B indicates the dew-point (about 6.5 °C), and if any air reaches a part of the structure just below this temperature, then condensation will occur. The chart also shows that air at 20 °C and 40 per cent rh contains about 6 g of moisture in 1 kg of dry air. If some internal activity adds a further 6 g say, by the drying of clothes – then the air will contain 12 g of moisture and be at a relative humidity of just over 80 per cent (point C). A drop in temperature to 17 °C will then suffice to allow condensation to take place (point D).

Any surfaces below the dew-point of the air immediately adjacent to them will suffer surface condensation. Visible condensation is seen frequently on

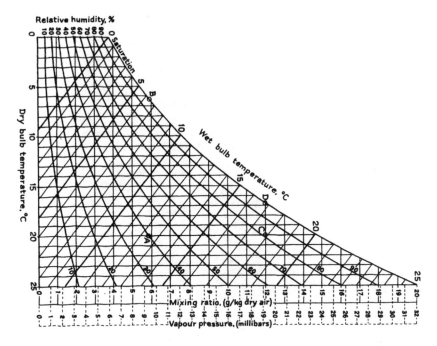

Figure 7.3 Psychrometric chart.

windows, external walls and on cold pipes and coldwater cisterns, and gives a warning sign that moisture levels are high. As vapour pressure is usually higher inside a building than outside, water vapour will move through a permeable structure towards the outside. In its passage, it may be cooled to the dew-point, causing condensation at some point within the constructional materials, in the spaces between them or at cold bridges, which are localised cool surfaces interrupting areas of better thermal insulation, for example lintels. Interstitial condensation, that is, condensation within the thickness of a material or structure, can be potentially harmful. A main effect of excessive moisture reaching the inside of a building – whether arising from moisture trapped during construction, as ground water, as direct rain penetration or from human activities – is to increase the relative humidity and the risk of condensation, and assist the growth of mould (Figure 7.4). Mould growth can occur if the rh remains above 70 per cent but, for active growth, prolonged spells of over 80 per cent rh are generally necessary or continued access to direct moisture supply within the material upon which they will form. The spores of fungi are ever present in the air in most buildings and so are nutrients; thus, the factor controlling growth is this supply of moisture. Condensation-related mould growths are usually dark in colour, often green or black, and both disfigure and cause deterioration

Figure 7.4 Severe mould growth.

of furnishings, fabrics, wallpapers and many decorative coatings. They can form on stored clothing and bedlinen, particularly when these are in unheated, unventilated built-in cupboards on external walls. Certain fungi can also rot timber and these are dealt with in Chapter 5. Suffice it to say here that heavy and repeated condensation running down windows and entering the timber frame through discontinuities in the paint film or in the putty can allow wet-rot fungi to develop. In the past, drained condensation channels were often provided, though condensation was not then a serious problem. Nowadays, when reduced ventilation rates and changes in living habits have made condensation a major problem, such drainage channels are rarities.

Condensation is the main source of the moisture needed for the growth of moulds on the indoor surfaces of residential buildings (Douglas 2006). Penetrating dampness above ground can also initiate mould growth on indoor surfaces, but its effects are usually localised. In contrast, ground-borne dampness, because it often contains salts such as chlorides or nitrates, normally does not encourage the growth of moulds on affected walls.

Mould, however, can cause significant health problems for those living in damp buildings (Committee on Damp Indoor Spaces & Health 2004; Howieson 2005). Common species of condensation-related mould are *Aspergillus fumigatus, Cladosporium cladosporioides Penicillium chrysogenum, and Stachybotrys chartarum*. For example, *Stachyborys chartarum* (sometimes called *S. atra* or "Black Mould") is a toxigenic mould that can trigger significant respiratory problems for those exposed to its spores – particularly infants and the elderly.

Moisture vapour reaching porous building materials will raise the moisture content of the air contained within the pores: if liquid water enters the material, it will directly raise its moisture content. The effect of both sorts of wetting is to increase the thermal conductivity of the material and so reduce its level of thermal insulation. If heating within the building remains constant, the effect will be to reduce the temperature and this reduction will itself help, still further, to increase condensation. (The effect may be, of course, to cause heating to be increased and to be manifested partly in greater fuel expenditure. If this heating is by unvented paraffin appliances or by unvented gas heaters, then still further amounts of moisture are generated.)

7.5   Avoidance of condensation

The amount of moisture available as a potential cause of condensation can be reduced by attention to the foregoing points. Thus, good design and construction can minimise the amounts of moisture entrapped during construction, entering from the ground and penetrating as rain or snow. Publicity and education may help in reducing the actual generation of moisture within a building, in promoting ventilation and increasing heating levels. It is probable, though, that little heed will be given to advice that results in higher fuel bills. Nevertheless, heating by oil and gas appliances which are not connected to flues should be actively discouraged. Mechanical extraction should be used to remove moisture from high emission areas, such as kitchens, bathrooms and drying cupboards. Electric dehumidifiers may have a role but they are not very effective in buildings where the main problem is low temperature rather than high vapour pressure. They are portable but rather noisy and will be more acceptable in kitchens than in bedrooms.

However, even when all such precautions are taken, it is inevitable that moisture will be present in buildings. Whether or not condensation will occur in a specific situation requires knowledge of many complex and interrelating factors. These include not only details of the design and the materials to be used but also knowledge of the occupations to be carried out in the building and their pattern.

Slight changes in habits can tilt the balance towards or away from condensation. In particular, much depends upon the levels of heating and thermal insulation used, ventilation and the permeability of the building fabric to moisture. Heating a building helps to raise the temperature of internal surfaces above the dew point of the air. The effect of increasing thermal insulation for any given heat input is to conserve the heat and thus make it more effective. The temperature of surfaces is also dependent upon the speed at which the building fabric heats or cools in response to changes in external and internal temperatures.

Ventilation assists in the removal of any moisture produced within the building. Natural ventilation is nearly always wholly beneficial in reducing the risk of condensation: modern building design, with its tendency to dispense with chimneys and flues, and the current trend to block chimneys in existing buildings has reduced these ventilation rates. The permeability of the construction influences the moisture conditions within the structure and can be changed greatly by the use of vapour control layers. The vapour resistance of some common membranes and materials is shown in Table 7.2.

Table 7.2 Vapour resistance of materials (Based on data contained in BS 5250: 2003)

Material	Vapour resistance, approximate [(MN s)/g]
Aluminium foil	Several thousand
Polyethylene: 10 μm	250–350
Gloss paint	7–40
Roofing felts	4.5–100
Stone: 100 mm	15–45
Brickwork: 225 mm	5–30
Expanded polystyrene: 50 mm	5–30
Bitumen-impregnated paper	11
Concrete: 100 mm	3–10
Wood wool: 50 mm	0.7–2
Rendering: 20 mm	2
Foamed urea-formaldehyde: 50 mm	1–1.5
Plasterboard: 9.5 mm	0.4–0.6
Kraft paper: 5-ply	0.6
Emulsion paint: 2 coats	0.2–0.6
Plaster: 5 mm	0.3

BS 5250 recommends design assumptions for dwellings relating to internal and external temperatures, to humidities and to ventilation rates. It recommends standards of thermal insulation which, taken in conjunction with those design assumptions, should prevent surface condensation in dwellings in all but the most exceptional circumstances. The principles used can be applied to specific floor, wall and roof constructions to predict the likelihood of condensation and the point in the structure at which it could occur. Concern with failures has so far related mainly to condensation which occurs on the inside surfaces of a building. Clearly, when this is sufficiently severe to cause damage to fittings, furnishings and decorations, such an emphasis is understandable and inevitable. However, condensation on windows and on internal surfaces and its effects are, at least, readily seen. As the standards for thermal insulation rise, it is probable that the point at which condensation occurs will be pushed back into the structure, towards the outside, where it may be hidden but do greater damage in the long run. It may not be possible to prevent interstitial condensation but it is necessary to ensure that it does not occur where it can cause the types of damage described in Chapters 5 and 6, particularly to timber and to metal fixings. Detailed knowledge of the thermal properties and behaviour of materials and structures under fluctuating temperature, moisture and ventilation conditions is still a field for research: guidance is available in the specialist literature to enable the risk and position of condensation to be assessed for a specific structure. If past experience is any guide, though, it would be unwise to believe that the extent and position of condensation can be closely controlled or that vapour barriers will prevent, in reality, all water vapour from passing. It will be wiser to assume that condensation will occur at any point within a structure and to ensure, by suitable design and specification, that the moisture can move freely and easily to the outside, either in liquid or vapour form, and that sensitive materials are desensitised, for example by preserving timber components, or replacing them by those products more resistant to moisture.

A modern solution to the problem of condensation in housing is positive input ventilation (PIV). Douglas (2006) describes the use of this technique in houses and flats.

7.6 Summary

The influence of moisture on the occurrence of defects cannot be underestimated. Over half of all defects will involve moisture in one form or another. Its main adverse consequence is dampness, which is still a troublesome problem in many buildings today.

Chapter 8

Foundations

OVERVIEW

This chapter focuses on substructure failures. It examines the main factors that affect the performance and durability of foundations. It also gives basic guidance on remedial measures to consider when faced with a substructure problem.

8.1 Introduction

It seems appropriate to start with foundations when dealing in detail with the avoidance of defects in specific building elements as that is where construction starts. In addition, serious defects in foundations are, generally, the most difficult and costly to remedy. The Building Regulations identifies three functions for the foundations of a building. The first is that they should sustain and transmit safely to the ground the various loads imposed by, and upon, a building, in a way which will not impair the stability of, or cause damage to, the building or adjoining buildings; second, their construction must safeguard the building against damage by physical forces generated in the subsoil; and third, they must resist adequately attack by chemical compounds present in the subsoil. Clearly, the ground or subsoil has a major influence on the ability of foundations to perform these functions, and its type, structure and properties are of prime importance.

8.2 Type and structure of the soil

Soils include materials of various origins but, for purposes of identification, two essential characteristics have been recognised – the size and nature of the particles composing the soil and the properties resulting from their arrangement (BS 5930). Five principal types can be distinguished. Gravels and sands are relatively coarse-grained, non-cohesive particles derived from the weathering of rocks. Gravels consist of particles which lie mostly

Table 8.1 Classification and characteristics of common soils

Type of soil	Size and nature of particle	Compressibility
Gravels	Mostly between 60 and 2 mm; non-cohesive.	Generally loose and uncompacted.
Sands	Mostly between 2 and 0.06 mm; non-cohesive when dry; may contain varying amounts of silt and clay.	Can be loose or somewhat compacted.
Silts	Mostly between 0.06 and 0.002 mm; some cohesion; little plasticity.	Capable of being moulded in the fingers.
Clays	Less than 0.002 mm; marked cohesion and plasticity; appreciable shrinkage on drying and swelling on wetting.	Very soft clays exude between the fingers when squeezed; firm clays can be moulded by strong pressure; stiff clays cannot be so moulded.
Peats	Fibrous organic materials; can be firm or spongy.	Have a high degree of compressibility.

between 76 and 2 mm: the particle size of sands lies between 2 and 0.06 mm. Within this range, sands may be classified as coarse, medium or fine. Silts are natural sediments of smaller particle size than sands, chiefly lying between 0.06 and 0.002 mm. Clays are formed from the weathering of rocks and have a particle size less than 0.002 mm. Peat is an accumulation of fibrous or spongy vegetable matter formed by the decay of plants. These five soil types and their characteristics are shown in Table 8.1.

8.3   Interaction between soils and buildings

Under natural conditions, water and air fill the spaces between the soil particles. The properties of soil are influenced greatly by the amount of water so held, and the volume and strength changes which may occur when this water is reduced or increased. Changes in the behaviour of soils influence that of the foundations in contact with them and this affects the behaviour of the superimposed building. The building itself, through the loads transmitted via the foundations, compresses the soil and can change its behaviour. Interactions between the soil, the foundations and the building are complex and highly dependent upon the forces involved when soils shrink or expand due to loss or gain of moisture.

8.4   Soil movement

8.4.1   Preamble

When the water present between soil particles is removed, the latter will tend to move closer together: conversely, when water is absorbed, they will tend to move apart. Large movement can occur with clays, for these are capable of absorbing and relinquishing large quantities of moisture: drying leads to shrinkage and a gain in strength, and absorption to swelling and a loss in strength. Movement in sands is for the most part negligible, for they have little capacity to hold water. Silts have movement which lies between that of clays and sands. Peat can exhibit very large movement and has little bearing capacity. Changes in water content of soils may be caused in several ways. The most obvious is that caused when the soil is loaded by the weight of the foundations and the superimposed building. Water is then squeezed out of the soil and the soil particles move closer together. As the ground is compressed or consolidated in this way, the foundations settle until equilibrium is achieved between the load imposed on the soil and the forces acting between its particles. The more clay there is contained in the soil, the longer does it take for this equilibrium to be achieved. With soils wholly of clay, such settlement may go on for years, while with sands it is rapid and is substantially finished by the time building is completed. It may be of interest to note that a reduction in loading, such as will be caused by demolition or excavation, can lead to water migrating towards the unloaded soil, causing it to swell – again, appreciable with clays and negligible with sands.

8.4.2 Effects of vegetation

Knowledge that movement can be caused by loss of water through the growth of vegetation and to gain of water by its removal seems to have been overlooked – at least, up until 1976, when the severe drought and hot weather in the United Kingdom led to a rash of troubles. In fact, problems associated with vegetation and climate were of long standing, and both researched and reported upon by Ward (1948) in the immediate post-war years. Tree roots can extract large quantities of water from the soil: a fully grown poplar uses over 50 000 litres in a year. When the soil is of clay, this will lead to a drying shrinkage, the magnitude of which will depend upon the inherent properties of the clay and, of course, on the nature of the tree and its moisture requirements. If tree roots take up moisture from under, or near to, foundations, the latter will subside and such subsidence will almost inevitably be uneven (Figure 8.1). The possible adverse effects on foundations and, thus, upon the final structure of the drying action of tree roots in areas of shrinkable clays is not appreciated by many involved in the construction process – and, unfortunately, by many owners of buildings. The small, immature tree, which seems to be a fair distance away initially, looks uncomfortably close in later years. The mature height of some common trees found in the United Kingdom is shown in Table 8.2. The distance to

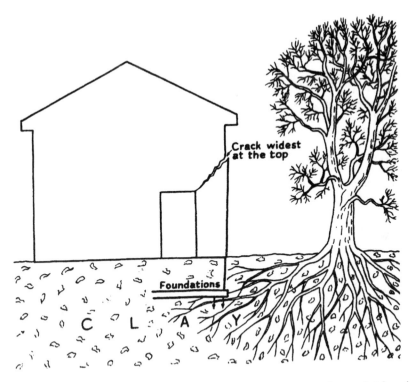

Figure 8.1 Effect of tree roots on foundations. Tree roots cause clay to shrink and foundations to drop.

which the roots of a tree spread depends largely upon the type of tree and its height. The roots of many common trees extend to a distance at least equal to their height. The roots of willow, elm and poplar can extend to twice the height.

It is also most important to understand that when trees are felled, clay soils will gradually swell as water returns to the ground. A clay site cleared of trees needs to be allowed to recover before building begins or, if this is not possible for economic or other reasons, then foundations need to be specially designed, as described later, to prevent damage caused by this swelling. Hedges and shrubs can also cause desiccation and shrinkage of clay and, if removed, can lead to swelling. While their effect is less than that produced by trees, it cannot be neglected for they are often grown closer to buildings. The clays which display the largest movement upon drying are the firm, shrinkable clays found particularly in South-East England, for example the London and Oxford clays, and the Weald and Gault clays.

Table 8.2 Heights of some common trees (based on data by Johnson 1999)

Common name	Approximate mature height (m)	Approximate height after 15 years (m)
Douglas fir	40	12
European larch	40	9
English elm	40	7
Lombardy poplar	30	9
European ash	30	7
London plane	30	5
White willow	25	15
Scots pine	25	10
Silver birch	25	9
European horse chestnut	25	9
Oak	25	8
Beech	25	7
Lime	25	7
Cedar	25	6
Weeping willow	20	9
English holly	15	4
Yew	10	5
Juniper	5	3

8.4.3 Other causes of ground movement

Major ground movement can occur over underground mining areas, as the ground collapses over the workings. Nowadays, good records are kept of mine workings but this was not so during much of the 19th century.

A rather uncommon combination of circumstances can lead to the expansion of ground when frozen. The soils mainly involved are silts, fine sand and chalk. In areas where the water table is high and when there are prolonged periods of freezing, ice lenses can be formed in these soils which cause heaving of the ground and the foundations upon it if they are less than 600 mm deep – a phenomenon known as frost heave. In practice, little trouble has been caused in the United Kingdom, and even in severe winters the ground at foundation base level near to occupied and heated buildings is not likely to freeze. The risk, albeit small anyway, is confined to unheated buildings and those under construction.

8.4.4 Effects of foundation movement

The greatest problems have occurred when shrinkable soils have dried excessively through the removal of moisture by nearby growing vegetation. Such drying is likely to be greatest at the corners of foundations. As the ground falls away, the weight of the building pushes the then suspended parts of the foundations down and the walls in that vicinity crack. Cracking is predominantly diagonal and follows the vertical and horizontal mortar joints in brickwork, unless the mortar is abnormally strong for the bricks used, when cracking may occur through the latter. The cracks are widest at the top corners of the building and decrease as they approach ground level. The appearance of cracks of this pattern at the end of an especially dry summer is a fairly sure sign of desiccation of a shrinkable clay soil. Door and window frames also distort due to the deformation of the walls, leading to their sticking or jamming. In severe cases, service pipes may fracture, walls may bulge and floors may slope noticeably. The cracks tend to close partly, following periods of prolonged rain, for example by the end of the following winter.

When trees, large shrubs and hedges are cut down before building, long-term swelling of clay soils can be substantial and can take place over several years. The upward forces on foundations can cause severe stresses at the corners of the building or may act more centrally. In the former case, cracking patterns are usually similar to those already mentioned but with the important difference that crack width is greatest near to foundation level and becomes narrower at the higher levels. When forces act more centrally, cracks tend to be straight rather than diagonal and are widest at the top. Often, there will be a single crack in each of the two opposite walls of the building and they may be connected with a crack in the floor if this is of concrete.

When subsidence occurs due to active mining operations, the building tilts towards the advanced working, and random forms of cracking generally occur. Diagnosis is fairly obvious through knowledge of the presence of active mining in the area. Over old forgotten workings, diagnosis is clearly more difficult but the presence of random cracking is a guide, though this can occur from other causes, for example from the use of poorly consolidated fill or from seismic tremors.

8.4.5 Avoidance of failure due to soil movement

Detailed guidance on the design and construction of foundations is given in BS 8004 and BS CP 101. The latter deals with non-industrial buildings of not more than four storeys and it is in this category of buildings that most post-war problems have occurred, particularly in housing. The National House-Building Council (NHBC) has produced a related manual intended for house builders, surveyors, engineers and architects (NHBC 2005), which gives a checklist of actions required to prevent trouble. A first need is for adequate site reconnaissance, and a study of any local recorded information, to determine the topography, the basic type of ground, the vegetation present, and the type and proximity of ground water, and to glean as much information as possible on the previous use of the site, for example any possible relationship to mining. Geological maps will give guidance on the main ground strata to be expected, but study of these needs to be augmented by the digging of trial pits (or hand-augered bore holes) and by descriptions in detail of the soil profiles revealed and the level of ground water. If such inspections reveal the likelihood of hazardous ground conditions, such as those associated with peat or shrinkable clay, or the presence of mining or of old buildings, then specialist advice needs to be sought. Failure to undertake these essentially simple steps has caused much expenditure in post-war years.

On sites where the soil is of firm shrinkable clay, it is necessary to take the foundation down to a depth which should eliminate significant ground movements. A foundation depth of 1 m is generally adequate in such circumstances, when the site is unaffected by trees. The NHBC has provided guidance on the depth of foundations required on sites with trees, relating such depth to the type of tree, its mature height, its distance from the foundations and its geographical location (NHBC 2005). Several types of foundation can be used in these circumstances and choice is likely to be dictated by financial considerations. The traditional strip foundation of concrete is usually some 150 mm thick with a minimum width of 450 mm, from which two leaves of brickwork are built up to DPC level. At a foundation depth of 1 m, at least twelve courses of brickwork would be needed. The cavity wall up to ground level is filled with concrete and the trench is backfilled with earth and hardcore. These traditional strip foundations are labour-intensive and as the depth required increases, trench fill or narrow strip foundations are being used increasingly. These are formed by cutting a trench narrower than that needed for the traditional strip and pouring concrete to a depth such that generally only four courses of brickwork are required to reach DPC level. It is important that setting-out is accurately accomplished to ensure that the brick courses do not oversail the edges of the narrower strip. As the depth for foundations increases, and in soils liable to swell, it can become unsatisfactory to use ever-deeper trench fill. Swelling pressures can act on the large areas of foundations then in contact with the soil to cause lateral movement and rotation of the foundations and also vertical movement on the sides of the trench fill. It will be better, and often cheaper, to use bored piles. The likelihood of damage from the swelling of clay soils following the cutting or removal of vegetation is difficult to predict but it is known that many years can elapse before movement can be considered complete. BS 8004 considers it unwise to assume that swelling will be completed within two winter seasons following the removal of trees and hedges, and calls for special foundations, such as bored piles and ground beams, to be used if buildings are sited over or close to former trees, shrubs or hedges. The timescale of swelling, and the distance over which the effect is appreciable, requires further research.

If additions are to be made to existing buildings in areas where soil movement is likely to be a problem, then it is probable that differential movement will occur between the new and old structure. It is necessary to ensure that such movement can take place without causing damage to either by the use, where possible, of flexible or sliding joints.

Where subsidence due to mining is a risk, expert advice needs to be sought on the design of foundations, which is outside the scope of this book. With the winding up of the National Coal Board in 1994, the relevant local council or the equivalent authority responsible for deep and other types of mining should be consulted. Detailed guidance on the types of foundation for low-rise buildings, with particular reference to special ground problems, including soils of firm shrinkable clay and mining subsidence, is given by Tomlinson, Driscoll and Burland (1978).

8.4.6 Remedial measures

Repairs to foundations are very expensive and, if things have gone wrong, much care should be taken in deciding whether repairs are necessary and, if so, the form they should take. There was a general over-reaction to damage caused by the 1976 drought and many unnecessary repairs were undertaken. Underpinning of foundations is an extreme measure to adopt and may do more harm than good in some circumstances, for example where swelling of the soil may take place afterwards through normal return of water to the soil in the wet season or following the wholesale removal

Table 8.3 Visible damage through foundation failure

Degree of damage	Description	Category of crack damage per BRE Digest 251
Very slight or slight	Fine cracks, not greater than 5 mm wide, often not visible in external brickwork and easily filled. Some slight sticking of doors and windows possible.	0–2
Moderate	Cracks may be typically from 5 to 15 mm wide; external brickwork will need repointing and some local replacement may be necessary. Doors and windows will stick and service pipes may fracture; general weathertightness may be impaired.	3
Severe/very severe	Cracks will typically exceed 15 mm in width and may exceed 25 mm. Walls are likely to lean or bulge noticeably and may require shoring; beams may lose their bearing. Window frames and door frames will distort and glass is likely to break. Service pipes are likely to be disrupted. External repair work will be necessary, involving partial or complete rebuilding.	4–5

of trees and vegetation at the time when shrinkage damage first became apparent. Table 8.3 shows a classification of visible damage to walls and the related ease of repair of plaster and brickwork.

8.5 Fill

8.5.1 Preamble

Good building land is scarce and this has put pressure on developers to fill other possible sites such as gravel pits, railway cuttings and open-cast mines. The support given by the fill depends crucially upon its type, the degree of consolidation it has reached and the way this has been achieved. All made-up ground should be treated as suspect because of the likelihood of extreme variability. The NHBC (2005) has recorded that the largest single cause of foundation failures to dwellings has been the use of poor fill. Mostly difficulties have arisen through settlement of the fill following inadequate compaction.

8.5.2 Settlement of fill

A long time is usually needed for the natural settlement of fill, particularly if the predominant particle size is small. Slow consolidation occurs, too, when the fill has been inadequately broken or graded and contains excessive voids. Considerable compaction of originally loosely compacted fill can occur later if water reaches it, perhaps through a rise in the water-table level. Sites containing domestic refuse are especially hazardous, for these contain materials which may have large voids, such as old metal, glass and plastic containers, and also vegetable matter which eventually decomposes and results in subsidence of the super-imposed foundation. Shrinkable clay used as fill will shrink on drying and can cause settlement difficulties, particularly if construction takes place when the clay is saturated during wet weather.

8.5.3 Heaving of fill

While settlement of fill is the major cause of trouble when building is on made-up ground, swelling of shale used as fill has also caused extensive damage, though the problem seems not to be widespread in the United Kingdom. Swelling shales are known to have caused failures in the USA and in Canada. In Britain, a series of failures occurred in the Teesside area and one in Glasgow in the 1970s. Investigations identified the ironstone shales which caused the trouble as belonging to the Whitbian shales of the Upper Lias, probably of the jet-rock series but also likely to be mixed with alum shales (Nixon 1978). The cause of swelling was attributed to the oxidation of pyrites in the shale, resulting in a marked volume increase. The oxidation process also produces sulphuric acid and this reacts with calcite present in the shale to form gypsum. The crystallisation of this gypsum between laminations in the rock is believed to be the predominant expansive force. The possibility of such problems occurring at points other than Teesside and Glasgow in the United Kingdom seems small but cannot be ruled out.

8.5.4 Effects of movement of fill

Movements due to settlement of fill are usually large and major cracking of external walls, screeds and internal partitions often results: doors and windows jam, gaps occur at heads of partitions and brickwork may bulge out. The expansive forces due to swelling pyritic shales cause concrete ground floors to lift and arch, which leads to cracking. Internal walls and partitions lift and crack, and there may be outward movement of the perimeter walls near to DPC level.

8.5.5 Avoidance of damage by fill

Wherever possible, it is better to avoid sites which have been filled and all possible information about the site should be obtained by visits there, by discussion with local people and the local authority and by studies of local maps. Site visits should aim at observing signs of damage to any buildings bordering the site and should include the digging of trial pits to assess the nature of the soil. If it is clear that the site has been filled and cannot be avoided, then numerous trial pits will be needed to assess the nature and variability of the fill and its boundaries, its depth, its chemical composition, the degree of compaction, and the method by which the fill seems to have been laid and compacted. In addition, the level of the water table will need to be monitored for at least a year before building begins, and the likelihood of further settlement should be assessed consequent upon the fill's becoming inundated. Adequate flexibility and protection to services to buildings will need to be provided. Specialist advice should be sought on ways of helping to consolidate the fill further and on the possible foundation solutions to the building to be erected by, for example, the use of piles.

Where chemical analysis has indicated the presence of pyrite and calcite, the best course to adopt is to remove the fill and replace it with a non-hazardous one. Where new fill is to be used on a site, the builder can, at least, exercise proper control over it. It should be of a granular nature, ideally a coarse sand or a gravel free from organic matter, and should be thoroughly compacted, layer by layer.

8.6 Summary

The importance of adequate foundations for all buildings cannot be overemphasised. However, care needs to be taken to avoid over-reacting to a perceived substructure problem. This could result in expensive and unnecessary remedial work such as underpinning being undertaken instead of basic repairs to the masonry affected.

Chapter 9

Floors, floor finishes and DPMs

OVERVIEW

This chapter outlines the problems associated with floors constructions. It also provides an insight into the defects affecting the associated parts of floors such as the substrate and finishes.

9.1 Background

Ground-bearing floors have not been a high risk area in recent years. However, failures have occurred, particularly in concrete screeds, by chemical attack, usually by sulphates, on the concrete base slab and through insufficient support to the slab by inadequately compacted hardcore.

9.2 Hardcore

9.2.1 Background

Hardcore is used to fill small depressions on sites and to adjust the amount of concrete needed in an over-site slab, following removal of topsoil from the site. It is also used on soft and wet sites to provide a good working surface and one which will not affect adversely the over-site concrete during placing. It has, too, some value in reducing moisture uptake from the ground. Hardcore is deemed to satisfy the Building Regulations (Section C4a) when it consists of clean clinker, broken brick or similar inert material free from water-soluble sulphates or other deleterious matter which might cause damage to the concrete. Materials mostly used in practice are concrete rubble, broken bricks and tiles, blast-furnace slag, various shales, pulverised fuel ash, quarry waste, chalk, gravel and crushed rock.

9.2.2 Hardcore and sulphate attack

A particular hazard to concrete floors can arise from the presence of soluble sulphates in hardcore or in the ground water. Solutions of sulphates can attack the set cement in concrete, as described in Chapter 6, the severity of attack much depending upon the type of sulphate present and the level of the water table. Broken bricks and tiles may contain soluble sulphates and, moreover, may be contaminated with gypsum plaster. Coal mining waste, too, usually contains soluble sulphates. Burnt colliery waste from old tips tends to have a higher soluble sulphate content than unburnt spoil and, when used as hardcore on wet sites, has frequently caused failure of over-site concrete. Other materials used as hardcore which can contain soluble sulphates include spent shales left as residue following the extraction of oil from oil shales, pulverised fuel ash, particularly if mixed with furnace bottom ash, blast-furnace slags derived from iron-making and shales containing pyrites (see Section 9.4.2).

Attack by sulphates on concrete floors, whether they derive from hardcore or from the soil, may be manifested initially by lifting of the floor and some binding of doors. As attack proceeds, major lifting and arching can occur, the concrete surface cracks and there may be some movement of the external walls near DPC level (Figure 9.1). These expansive forces are usually slow

Figure 9.1 Sulphate attack on concrete floor.

to develop and movements may not become apparent for several years. Such sulphate attack may be avoided or greatly reduced in severity by using, for hardcore, materials such as coarse sand, gravel, crushed rock, clean concrete rubble and quarry waste, which are usually free from soluble sulphates; by ensuring that the concrete placed is of low permeability; and, if there is still a possibility of attack, by using a type of cement low in tricalcium aluminate, such as sulphate-resisting Portland cement complying with BS 4027. An estimate should be made of the sulphate content in samples of ground water. BRE Special Digest 1 gives guidance on the selection of the type of cement and quality of concrete required to resist attack by sulphate-bearing ground waters and soils. It also provides information on the method for sampling and analysis of ground waters. Where soluble sulphates are present in hardcore, a water barrier such as polyethylene sheet at least 0.2 mm thick should be placed to separate the floor slab from the hardcore.

9.2.3 Hardcore and swelling

Similar damage can be caused to floors by the swelling of hardcore due to factors other than sulphate formation. Materials likely to swell from these causes are principally slags derived from steel-making, for these may include unhydrated lime or magnesia, some colliery spoils containing clay and refractory bricks used in chimneys and furnaces. These materials therefore should not be used as hardcore.

9.2.4 Hardcore and compaction

If hardcore is not thoroughly compacted, it will consolidate after the building has been completed and a solid ground-floor slab will no longer be

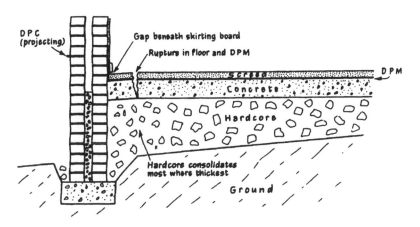

Figure 9.2 Floor slabs may settle and crack if hardcore thickness is excessive.

supported adequately over its whole area. This is particularly likely to happen when the depth of hardcore used is excessive as may occur on a sloping site, or where deep trench-fill has been used for foundations. Thorough compaction then becomes, if not impossible, at least unlikely. The solid slab is then likely to drop and crack particularly towards the edges under which the depth of hardcore is greatest. The first sign of trouble is usually the appearance of gaps between the floor and the skirting board. Cracks may appear in partitions. To avoid trouble, hardcore should be as well compacted as possible and a suspended floor, not a solid slab, should be used if the thickness of hardcore anywhere exceeds 600 mm (Figure 9.2).

9.3 Damp-proofing of floors

To prevent the rising of moisture from the ground and into the floor finish, it is necessary to provide damp-proofing both in the walls and in solid ground floors. A horizontal DPC in walls is required to be not less than 150 mm above ground level. Concrete bases and screeds will allow ground moisture to pass through and it is necessary to provide a damp-proof membrane (DPM) either as a sandwich layer within the thickness of the concrete or upon the surface of the concrete. (In the latter case, the damp-proof layer usually provides the final floor finish, for example pitch mastic or mastic asphalt.) These two requirements seem to be generally understood, though not always properly specified and achieved. A crucial need, however, is to link the DPC in the wall with that in the floor by means of a projection of the DPM (Figure 9.3). This is often not done, possibly because the designer

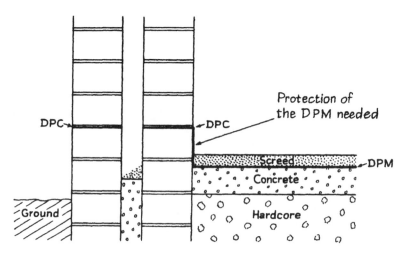

Figure 9.3 Continuity between wall DPC and floor DPM. This is needed but often omitted.

intended that this linkage should be provided but did not communicate the intention to the builder, feeling it to be a matter of normal good practice. Sometimes linkage has been made impossible by the power-float, used on the slab cast above the DPM, cutting off the projecting DPM. Whatever the reason, damp penetration has occurred in this manner in many post-war houses, a common effect being to cause decay of timber-skirting boards. Many materials may be used to provide damp-proofing for floors and BS 8102 provides information on their limitations. It also gives an indication of the properties of common flooring materials in relation to resistance to ground-moisture penetration. It may, however, be worth making the special point that cold applications of bituminous solutions, coal-tar pitch/rubber emulsions and bitumen/rubber emulsions will only give an impervious membrane if an adequate thickness is built up throughout. This will mean at least 0.6 mm and this will need two or more applications conscientiously brushed on – not easy to check after the event. Such cold applications must not be diluted. If polyethylene sheet is to be the DPM, then it must have a minimum nominal thickness of 250 μm (0.25 mm) and a minimum spot thickness at any point of 200 μm and the laps between sheets must be double-welted. Special care and detailing will be needed to ensure the continuity of any DPM if this is punctured by services.

9.4 Concrete floors

The design and construction of concrete floors, including granolithic and terrazzo, is dealt with in BS 8204. The main problem with concrete floors, apart from sulphate attack, has occurred through failure to remember that concrete shrinks on drying and, in so doing, tends to crack and to exert stress at the interface between it and floor screeds and finishes. There has also been a failure to recognise that concrete does not present a surface which allows ready bonding of superimposed finishes.

9.4.1 Screeds

Most constructional problems have occurred with concrete floor screeds. It may well prove possible, however, to make the surface of a concrete base sufficiently level and smooth to accept the final floor finish without using a screed, and this has much to recommend it. If a screed has to be applied, care is needed to get a good bond between it and the concrete base. If the bond is poor and shrinkage stresses are high, the screed will crack, with a tendency to curl at the edges of the cracks: if the screed is tapped the floor sounds hollow. Finishes applied over the concrete screed, such as tiles and sheet coverings, are also likely to crack and split and, ultimately, lose their adhesion to the screed (Figure 9.4). There are several factors which enhance the shrinkage and cracking of concrete screeds and weaken its bond to the

Figure 9.4 Curling and fracture of plastic tiles.

base concrete. The principal ones are inadequacies in the mix design of the screed and of the base on which it is to be cast; poor texture of the surface of the base concrete; too long an interval between casting the base concrete and the screed; an inadequate curing regime for the screed; too large an area of screed laid in one operation; and too thick a screed.

The mix proportions of the screed should be such that the ratio of cement to aggregate should not be greater than 1:3 nor less than 1:4.5 by weight of the dry materials. Suitable aggregates and the British Standards to which they should conform are listed in BS 8204. The amount of water used should be just sufficient to allow the screed to be properly compacted. As water content increases so does screed shrinkage on drying and the amount of laitance which, when dry, shows as excessive dusting and crazing of the surface. The base concrete should have mix proportions not leaner than 1:2:4 of cement:dry fine aggregate:dry coarse aggregate by weight and should be thoroughly compacted to provide a firm and strong foundation. A common cause of failure is a poor condition of the surface of the concrete base, leading to inadequate bonding of the screed. The texture required for the surface of the base depends partly upon the time interval between casting the base and the screed. If the screed is to be placed before the concrete base has set, all that is needed is for the surface of the base to be swept to remove cement laitance, water and any other material, such as leaves, which may lie upon it. Complete bonding should then be successfully obtained by this form of monolithic construction if the screed is laid within 3 hours of casting the base. Even though the screed and the base have different shrinkage characteristics, both can shrink together, which greatly reduces the tendency of the screed to curl and crack. The likelihood of the shrinkage forces exerted by the screed causing problems can be further minimised by ensuring that its thickness does not exceed 25 mm. It should, however, not be less than 10 mm thick. If a screed has to be placed, the monolithic form of construction is likely to give the greatest chance of success and should be used where possible and, particularly, if aggregates are used which are themselves liable to shrinkage (see Chapter 6). When monolithic construction is not possible, and the screed needs to be applied on a set concrete base, more rigorous treatment of the surface of the latter becomes necessary. This should be brushed with a stiff broom before it finally hardens, to remove laitance and loose aggregate, and to roughen the surface. Older concrete bases will need hacking and cleaning. The base concrete should be well wetted to reduce suction, preferably overnight, and a grout of cement and water brushed on, keeping just ahead of the application of the screed. A minimum screed thickness of 40 mm is needed. Where the screed is to be placed over a DPM, then there can be no bonding with the base and the minimum screed thickness will need to be 50 mm at all points. A mechanical compactor or a heavy tamper must be used to obtain satisfactory compaction of the screed. The ordinary hand float will not give the necessary consolidation, no matter how hard it is wielded. Screeds of thickness greater than 25 mm should be compacted in two layers one immediately after the other.

Floating screeds are those laid over a compressible layer of thermal or sound-insulating material. They are particularly prone to crushing and to cracking at gaps left between the individual slabs comprising the compressible layer. At such gaps, the screed can enter and form a solid bridge with the concrete base. To avoid crushing, floating screeds should be at least 65 mm thick or, if it contains heating elements, at least 75 mm thick (Figure 9.5). Penetration of the screed into joints can be prevented by laying impervious sheeting such as polyethylene to form a continuous layer over the compressible material. It is desirable to provide 20–50 mm mesh wire netting laid directly onto the sheeting to protect it, and the thermal or sound-insulating material, from mechanical damage when the screed is placed. To ensure that the compressible layer remains fully effective for its purpose, it should be turned up at the perimeter of the floor. The screed generally needs to be trowelled to provide the smooth, dense surface required to take thin floor finishes, such as PVC tiles. Excessive trowelling can cause the surface to craze and, when dry, to dust. The effort required to produce this defect,

Figure 9.5 Floating screed may crush if too thin.

however, suggests that this is one type of failure likely to decline with the passing years.

If large areas of screed are placed in one operation, random cracking often occurs. This can be reduced by laying screeds in separate bays and was, indeed, the recommendation in the past. However, it proved difficult to prevent curling and slight unevenness at the junctions of bays and the resulting waviness showed through thin floor finishes. Not only did such undulations spoil the appearance of large areas of flooring but cracking of the thin finishes occurred. Remedial work required the removal of large areas of floor finish around the bay junction and the grinding down of the uneven edges, a costly procedure. It is better not to lay screeds in bays when thin floor finishes are to be applied. Dense screeds containing heating elements need special care to reduce the effects of shrinkage, and laying in bays becomes necessary and specialist advice will need to be sought.

If concrete screeds dry rapidly, the risk and extent of cracking are increased, for the concrete is likely to lack strength sufficient to withstand the shrinkage forces. A simple protection, by covering with polyethylene sheeting or similar material for at least 7 days, and then allowing the screed to dry naturally, is all that is needed, but this is frequently neglected. The operation of screed laying and finishing is one where conscientious work-manship on site is vital for success, but this is often lacking and has been the cause of many failures.

If poor adhesion between screed and base is suspected, and confirmed by the hollow sound generated when the former is tapped with a hammer, it does not follow that costly remedial measures are necessary. If hollowness is fairly local, it may not matter. Repair may only be necessary if the hollowness is accompanied by visible lifting of the screed, such that it is likely to break under the superimposed loads it is designed to take.

9.4.2 Granolithic concrete and terrazzo

Granolithic concrete is concrete suitable for use as a wearing surface and is made with aggregates specially selected to provide the surface hardness and texture required. It is used principally for factory floors and does not have any further finishing material laid upon it. The difficulties that have occurred with granolithic concrete are similar to those with screeds, namely, shrinkage cracking, poor bond to the base concrete and dusting of the surface. These difficulties are minimised in the ways already outlined in Section 9.4.1, with, once again, monolithic construction offering the best chance of success. To reduce curling to an acceptable amount, and because no thin floor finishes are to be applied, BS 8204 recommends construction in bays, the maximum sizes of which depend upon the thickness of the concrete base and whether the granolithic concrete is to be laid in one monolithic operation. If it is, then maximum bay sizes should not exceed $30\,m^2$ where the thickness of the base concrete is at least $150\,mm$, or $15\,m^2$ for a $100\,mm$ thick base. This smaller bay size is also recommended when the granolithic finish has to be laid separately on a set and hardened base. For success, granolithic concrete requires a high level of skill and close supervision and, unless these can be guaranteed, it is better avoided. Terrazzo consists of a mix of cement and a decorative aggregate, usually marble, of a minimum size of $3\,mm$, and this is laid on a concrete screed. The failures that have occurred are of cracking and crazing, and lack of good bond to the screed. The risk of surface crazing is increased by too rapid drying. This has been a common problem caused by failure to cover the terrazzo with sheeting and by too great an absorption of water from the terrazzo mix into the screed. Surface crazing can be caused, also, by too rich a mix but the use of an excessive amount of cement seems hardly likely to be a common problem today. Crazing and cracking may mar the decorative appearance of both granolithic concrete and terrazzo but are usually not otherwise serious. They are better tolerated than repaired, for this, particularly for terrazzo, is likely to be an expensive and specialised operation.

9.4.3 Shrinkage of suspended reinforced concrete floors

Suspended concrete floors usually deflect slightly and cause little trouble in so doing. Occasionally, greater deflection can occur, leading to cracking at the base of partitions because these are not then properly supported by the floor. This may, of course, be due to inadequate structural design or construction but the possibility that the cause is the use of shrinkable aggregates (see Chapter 6) should also be considered, especially if such a failure occurs in Scotland.

9.4.4   Other forms of decay of concrete floors

Concrete is resistant to attack by the types and amounts of chemical likely to be used in a domestic environment, but attack by acids, vegetable oils, fats, milk and sugar solutions can occur in the related industrial environments: mineral oils and greases are not troublesome. Whether or not attack is likely, much depends upon the specific factory operations and each case must be assessed separately. There is a need, however, to protect the structural concrete by the use of an impermeable membrane, such as asphalt, bituminous felt or plastic sheeting, and to provide adequate drainage to the floor to facilitate rapid removal of spilled material.

9.5   Magnesite flooring

A floor finish not commonly used, but not yet extinct, consists of a mixture of calcined magnesite, together with a range of organic and inorganic fillers and pigments, gauged with a solution of magnesium chloride. This brew goes by the name of magnesium oxychloride or magnesite. The biggest problem arises because magnesium oxychloride is particularly susceptible to moisture and deteriorates when exposed to damp conditions. These conditions may arise if water penetrates the finish from below, perhaps, because of the omission of a DPM or the use of an inadequate one or, from above, from cleaning water or through condensation. When moisture attacks magnesium oxychloride, the surface and, indeed, the main body of the finish disintegrates. Magnesium oxychloride floors have an inherent tendency to sweat, since magnesium chloride takes up moisture from the air. Beads of moisture appear on the surface, much as would occur through condensation, and sweating can be confused with the latter. Metalwork, for example plumbing and electrical services, in contact with magnesium oxychloride, is liable to corrode even when the floor is dry.

The durability of magnesium oxychloride flooring and its resistance to cracking and sweating are very dependent upon the mix proportions and upon the construction technique used. A common error is to use excessive amounts of the magnesium chloride solution with which to gauge the dry ingredients; another is failure to allow the floor to harden undisturbed for at least 3 days. Failure to maintain the surface by a good wax polish also enhances the risk of sweating, and excessive use of cleaning water, particularly if it is used with strong alkalis such as soda, will lead to gradual disintegration. This flooring needs very careful design and conscientious workmanship to perform successfully and guidance is given in BS 8204 which, inter alia, requires that metalwork should be isolated from the floor finish at all times by at least 25 mm of uncracked dense concrete or by a coating of bitumen or coal-tar composition.

Once a magnesium oxychloride floor has started to disintegrate, it will be necessary, in general, to replace it. Sweating may possibly be overcome by improving ventilation, together with repeated applications of a good wax polish.

9.6   Timber flooring

Recommendations for laying timber flooring are contained in BS 8201. Moisture poses the main problem, for this can cause decay and movement. The causes and effects of dry rot and wet rot are dealt with in Chapter 5 and general ways of minimizing the amounts of moisture present in Chapter 7. Specifically, timber floors, whether of board, strip, block or mosaic, need to be kept from becoming damp.

Timber in contact with a concrete base needs to be protected by a damp-proof layer, either a surface or a sandwich membrane. Suitable materials are listed in BS 8201. Failures have occurred not so much through omission of such membranes but more because of inadequate jointing with the wall DPC. The DPM in the floor needs to be continuous with the DPC in the wall. When a sandwich membrane is used, it is necessary to allow sufficient time for the superimposed screed to dry before laying the timber finish upon it.

The length of time required depends much upon individual circumstances, but BS 8201 gives a rough guidance that one should allow 1 month for every 25 mm of thickness of screed above the DPM. It also recommends that, before laying the wood floor, the state of the concrete base should be checked by a hygrometer or other reliable method. Relative humidity readings up to 80 per cent indicate suitability to receive wood flooring.

In recent years, cases have been reported where dry rot has attacked timber floors and skirtings through infection from hardcore containing pieces of decayed timber. As stated in Chapter 5, the dry rot fungus is adept at spreading and penetrating cracks, and these can occur in the site concrete when hardcore settles under load. If dry rot does occur through infection from timber in hardcore, eradication is difficult and costly. Prevention is greatly to be preferred, by thorough inspection of hardcore and the removal of any wood in it before use.

Suspended timber ground floors need to be adequately cross-ventilated by providing air-bricks to give at least $3000 \, mm^2$ of open area per metre run of external wall. Building owners and, in particular, keen gardeners should be careful not to obstruct the airbricks in any way such as, for example, by the construction of an external patio. Good practice requires the provision of over-site concrete or other damp-resisting material to restrict the passage of water vapour from the ground into the space below the floor boards. It must be regarded as poor practice merely to remove topsoil from within the perimeter walls. DPCs are, of course, required in all sleeper walls.

The movement of timber as it absorbs and loses moisture is described in Chapter 6. A common defect with wood-block flooring is due to expansion of the blocks through uptake of moisture. When the forces associated with this expansion can no longer be accommodated by compression of the blocks, the latter will usually lift and arch, either separately from the adhesive which fixed them to the base concrete or by pulling the adhesive away from the concrete. The position of lifting may occur anywhere but is most common at the perimeter of the floor. Probably, the most common cause of moisture penetration in wood blocks is through failure to provide an adequate DPM under the blocks: too frequently, the adhesive used to stick them is taken, misguidedly, as providing an adequate barrier. Hot-applied bitumen or pitch used as the adhesive will give a proper barrier but single applications of cold bituminous emulsions will not. It is important, in order to minimise potential movement, that wood blocks and, indeed, wood flooring of all types are laid at a moisture content likely to be close to that encountered in service. It is also necessary to provide a compression joint, for example, of cork, around the perimeter of the floor. If arching has occurred through dampness of a temporary nature, most commonly when the moisture in the blocks comes from construction water in the concrete screed, it will usually be possible to re-lay the blocks, when they, and the screed, have been given time to dry (Figure 9.6). The blocks should, of course, be examined for any signs of fungal attack before the decision is made to re-lay them. A less common defect is that caused by using blocks at too high an initial moisture content. When shrinkage occurs, gaps develop between blocks which spoil the appearance of the finished floor: there may also be some curling.

Joist hangers are commonly used to support timber joists but are often inadequately fixed principally by not bearing directly onto level masonry and by the back of the hanger not being tight against the face of the masonry. The joists used to support the overlying suspended timber floor may also be too short to sit properly within the hanger. In such cases, the masonry into which the hanger has been built can be over-stressed leading to local crushing, particularly where low-strength blockwork has been used for the inner leaf. Additionally, the hanger may move under load and the floor can settle, becoming uneven and springy. Plaster around the perimeter of the ceiling may also crack. It is important that hangers are bedded directly onto level masonry without any packing, that the backs of hangers are tight to

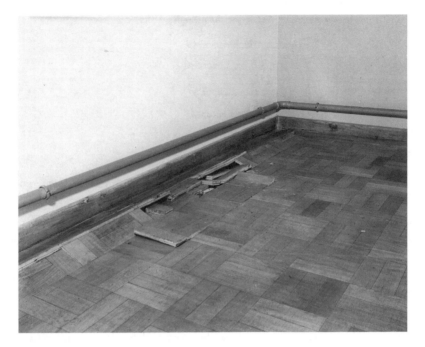

Figure 9.6 Arching of wood blocks.

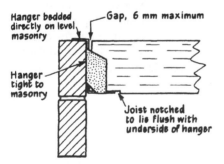

Figure 9.7 Joist hanger and suspended timber floor: good practice giving firm and level support.

the inner face of the blockwork, and that the gap between the end of the joist and the back plate of the hanger does not exceed 6 mm (Figure 9.7).

Hangers used should match joist sizes and must be of a grade consistent with the masonry. They should never be used on masonry other than that for which they are designed (Figure 9.8). Joists should be notched out to sit

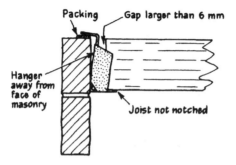

Figure 9.8 Joist hanger and suspended timber floor: bad practice leading to local crushing of masonry and uneven floors.

flush with the underside of the hanger so that plasterboard for the ceiling can be fixed properly without distortion.

9.7   Clay floor tiles

Clay tiles suitable as floor finishes are described in BS EN 14411. The failures which have occurred have been caused by differential movement between the base concrete or screed and the tile finish. Burnt clay and concrete have somewhat different coefficients of thermal expansion (see Table 5.4) which lead to different amounts of movement with temperature changes. Freshly laid concrete bases or screeds shrink appreciably as they dry, and this drying shrinkage may well be opposed to moisture expansion of the clay tiles, which occurs as moisture is absorbed by the tile. The moisture may be derived from the air, from cleaning water, from the ground if there are defects in damp-proofing (Section 9.2) and from the screed itself. This irreversible expansion of burnt-clay products is described in Chapter 6 and, while most of it occurs early in the life of the tile, later residual movements, if unable to be accommodated, can cause stresses to build up. Failures due to contraction of the concrete and expansion of the tile are manifested by arching of the tiles, often over a large area, or the forming of ridges, usually over one or two rows. In so doing, the tiles separate comparatively cleanly from the bedding and there is no general disintegration of either the tiles or the screed. Clearly, the best way to avoid failure is to minimise the movements which occur in the concrete screed or base and the tile, and to prevent direct transference of the stresses from one to the other. The worst conditions would occur if newly fired tiles were firmly bonded to a new concrete screed, for maximum moisture expansion and drying contraction movements could be expected. Tiles should not be used fresh from the kiln – even ageing of a fortnight will greatly reduce the irreversible moisture expansion. Direct bonding between tile and base should be prevented, and the best way of doing this is described in BS 5385. The tiles are bedded in mortar onto a separating layer such as building paper, bitumen felt or polyethylene: sand should not be used as such a layer. These separating layers have the additional advantage of giving some resistance to the passage of moisture from the base into the tile. However, even with their use, movement joints are needed around the perimeter of the tiled area. Intermediate joints are needed in large areas and at points where stresses in the concrete base are most likely to occur, for example over walls or beams in the case of tiled suspended floors. The concrete base and screed should be fully mature before tile laying begins and the floor should be closed to all traffic for 4 days after completion of tiling and then only foot traffic permitted for a further 10 days.

9.8 Plastic sheets and tiles

Sheet and tile flooring made from thermoplastic binders (principally PVC) are applied mostly to concrete screeds and the commonest cause of failure is through moisture passing from the screed. Such water contains alkalis derived from the concrete and these chemicals attack the adhesives used to stick the sheet or tile flooring. The adhesive becomes detached from the concrete surface, though it often remains well attached to the plastics flooring. The sheet or tile becomes loosened, the edges lift and damage may then be caused to them by normal traffic. Degraded adhesive, oozing through the joints, may also cause staining. Sodium carbonate can be left, following evaporation of water which has penetrated the joints in thermoplastic and vinyl asbestos tiles and, commonly, forms white bands around the edges of the tiles. Such a deposit will not, in many cases, prove harmful though it mars appearance. However, under particularly damp conditions, the salts may be absorbed into the body of the tiles and, on crystallising, cause degradation of the edges.

The main reason for attack is through leaving insufficient time for the screed to dry before applying the sheet or tile and, at least, can be remedied easily. If, however, the DPM in the concrete floor is ineffective or absent, rising damp will continue to act adversely until the fault is remedied. With plastics sheet flooring, rising dampness through a defective DPM can also cause areas to become loose and to blister or ripple. A defective junction between the DPC in the wall and the DPM in the floor is likely to be the trouble if failures are localised near external walls, though this is not invariably so, for failures can occur some distance away from the entry point of the moisture. Excessive use of cleaning water can also cause loss of adhesion and some blistering.

Sometimes, inadequate bond between adhesive and tile can occur even when conditions are dry. This is caused by too long a delay between spreading the adhesive and laying the tile, leading to loss of surface tack. This is a fault of workmanship and is manifested principally by the looseness of individual tiles and an absence of staining. Some adhesives can cause flexible PVC flooring to shrink through migration of the plasticiser into the adhesive, the latter being softened as a consequence. Under traffic, large gaps can then appear between tiles. This can happen, too, if the solvent contained in the adhesive has become entrapped through too early an application of large areas of sheeting.

Most failures with plastics sheet and tile flooring can be prevented if care is taken to ensure

- that there is an effective DPM;
- that the screed is allowed to dry properly before application;
- that the use of cleaning water is kept to a minimum;
- that the finish is applied neither too soon nor too long after application of the adhesive; and
- that only those adhesives specifically recommended by the manufacturer of the tile or sheet finish are used.

The repair that will be feasible should failures happen will much depend upon the cause. It may be possible to re-use loosened tiles or sheeting if all that is needed is to give more time for the screed to dry. The absence or defectiveness of a DPM, however, may involve more radical repair which, in turn, may necessitate a change in the type of floor finish used.

9.9 Summary

Of all the elements in and on a building, the floors are subjected to the most wear and tear and maintenance treatment. They are also vulnerable to aggressive agents from below (ground moisture and soil chemicals) as well as chemicals and imposed loadings from above. As a result, floors, floor finishes and DPMs have to be especially resilient to prevent premature failure.

Chapter 10

Walls and DPCs

OVERVIEW

This chapter deals with the many problems affecting walls. It addresses the issues of stability and weathertightness, both of which are common walling problems.

10.1 Moisture penetration from the ground

Rising dampness in walls is likely to cause damage to internal plaster and decorations, particularly when hygroscopic ground salts are brought up in solution, as they generally are. The line of dampness due to the complete absence of a DPC is usually fairly continuous and roughly horizontal, and can extend several feet in bad cases.

Remedial measures much depend upon the nature of the wall. If they are of brickwork or coursed stonework, in some instances, it may be feasible to cut a slot in the wall and insert a DPC or to inject a chemical damp-proofing system. Old walls of random masonry construction are unlikely to yield to these treatments and these can be nearly impossible to damp-proof. But see Douglas (2006) for measures to minimise the need for a retrofit DPC.

However, this book is concerned principally with the avoidance of failure in new construction and one would not expect the complete absence of a DPC in such work. A more common problem in modern construction is bridging of the cavity wall by excessive mortar droppings at the bottom of the wall (Figure 10.1). This leads to irregular damp patches appearing adjacent to the mortar droppings. If such poor workmanship cannot be prevented then, fortunately, the remedy is not too expensive (see Figure 10.2). Bricks will need to be cut away and the cavity cleaned out. Inadequate laps in flexible DPCs can also lead to moisture penetration. Laps should be of at least 100 mm and the DPC laid on a full mortar bed and also fully covered with mortar above to prevent damage. It is not uncommon for the DPC in the external wall to have been bridged by pointing,

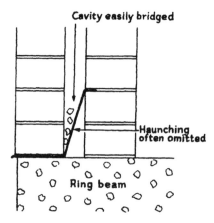

Figure 10.1 Bridging of wall DPC. DPC is seldom supported; mortar droppings are difficult to remove.

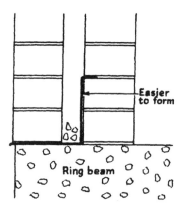

Figure 10.2 Preferred wall DPC. DPC needs no haunching; more droppings are necessary before DPC is bridged.

or rendering, over it. This should be prevented by projecting the DPC slightly through the external face and by stopping any rendering short of it. Bridging can also occur through over-zealous gardening, a common cause being the heaping of soil against the wall or the construction of patio paving above DPC level. Such gardening activities, if afterwards remedied, will give the home owner two lots of backache and a stained wall.

10.2 Rain penetration

10.2.1 Background

Eaves, canopies, cornices, string courses and other forms of overhang have been used traditionally to protect walls from rain and its effects but in post-war years their use has declined. Flat roofs have become more frequent and, where pitched roofs still prevail, eaves overhangs have diminished. The extent to which such features affect the degree of wetness of walls, and the time for which they remain wet, has not been adequately researched. Though overhangs can affect the air stream, particularly in tall buildings, with uncertain effects, it is probable that walls, in general, are wetter for longer when they are absent.

The ability of a wall to exclude rain thus depends partly upon such design features but principally upon its topographical and geographical situation. Table 10 of BS 5628 Part 3 gives exposure categories either in terms of the local spell indices calculated using BS DD 93 or those based on the BRE driving rain index (Lacy 1976). It also depends greatly upon the materials and construction of the wall. Generally speaking, rain is likely to penetrate solid brickwork or blockwork which is not rendered, usually directly through cracks in the mortar, and between mortar and brick or block. The risk is increased if the mortar is of poor quality and if the vertical joints of masonry are filled inadequately (e.g. because of "tip pointing"). Rain penetration shows as damp patches on the internal face of the wall usually within a few hours of rain falling: when the wall dries, a stain is often left. Solid masonry walls are now seldom built, and little more need be stated here, except to warn against some possible difficulties if water-repellent solutions are re-applied to reduce the risk of rain penetration in existing walls. The trouble with such coatings is that they retard considerably the subsequent evaporation of any water that may get in through lack of continuity in the coating or through unfilled cracks and, in so doing, can lead to more persistent dampness. To be successful, a continuous, crack-free coating is needed, which may be difficult to achieve. When soluble salts are present in a wall, they can, when the wall remains untreated, move in solution to the outer face and, under dry conditions, appear on the surface as an efflorescence. A water repellent may prevent this free movement, and salts can be deposited within the pores of the masonry and some distance beneath the surface. When they crystallise on drying, the forces produced can cause spalling of the surface. A preferred treatment, though admittedly one which changes the appearance of the wall, is to clad it with a porous rendering or with tiles. Cavity walls are now used for the great majority of buildings and should, in theory, prevent the direct penetration of rain to the inner leaf. In general, they do, but plenty of problems have occurred through bridging of the cavity by mortar droppings; by poor positioning of wall ties; by unwise use of cavity fill for thermal insulation; and by inadequate design and construction of DPCs at junctions. It is worth making the point that rain will penetrate the outer leaf of most walls and run down the inner face of that leaf. Penetration is likely to develop quickly and to an appreciable extent in some types of wall, particularly concrete masonry block walls. BS 5628 Part 3 calls for any mortar which unavoidably falls on the wall ties and cavity trays to be removed daily. This is not a counsel of perfection but a statement of what used to be a normal brick-laying operation. If it is not done, damp patches are likely to show on the inner leaf within a few hours of rainfall. Such patches are likely to occur sporadically and the use of a metal detector will, most probably, indicate that a tie is close by. Remedial measures usually involve taking bricks from the outer leaf, just above the site of the dampness, and removing the mortar, or even sometimes broken bricks, which have been dropped down the cavity. Ties should not be bedded with a fall from the outer to the inner leaf but sometimes are, and, in these cases, can allow rain water to trickle down the tie to the inner leaf. The fault can be due to failure to match courses by bringing up one leaf of the wall too far in height above the other during construction. Remedial measures will necessitate the rebedding of the ties but whether this will prove possible will much depend upon the degree of mismatch of the mortar joints. Ties are, not infrequently, poorly positioned laterally so that the central projection in the tie, designed to shed any water penetrating the outer leaf, is not in the centre of the cavity but close to, or touching, the inner leaf (Figure 10.3). Fortunately these faults, or at least the manifestation of them, are not as common as the lodgement of mortar droppings, for they are inherently more difficult and expensive to put right.

10.2.2   Cavity fill for existing buildings

In recent years, the need to conserve energy has led to the provision of thermal insulation within the cavity of a cavity wall. For existing buildings, a commonly used material was urea-formaldehyde, foamed on site and injected into the cavity to fill it. As the main purpose of the cavity is to prevent rain penetration, the filling of it clearly, in principle, increases the risk. Whether or not rain penetration to the inner leaf will occur depends upon many factors, of which the main one is the adequacy of the external wall against wind-driven rain.

Rain can pass through fissures in the foam. These can arise when urea-formaldehyde sets and shrinks. The extent to which such fissuring can occur depends principally upon the degree of control exercised during injection and upon the precise formulation of the foam. Further risks of penetration arise if the upper boundary of the foam is not fully protected from the ingress of rain. There have been failures due to the operation of one or more of these factors, though numbers so far have not been great. When they occur to the extent which leads to visible signs on the inner leaf, these

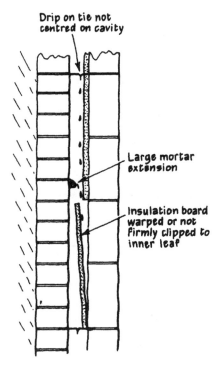

Drip on tie not centred on cavity

Large mortar extension

Insulation board warped or not firmly clipped to inner leaf

Figure 10.3 Bad practice: tie not centred; mortar extruded; board warped or poorly fixed. Rain is fed to inner leaf.

take the form of patches of damp associated with periods of rainfall. They cannot readily be distinguished from rain penetration caused by dirty wall ties. The risk will be minimised if, before filling the cavity, the suitability of the materials and construction of the external leaf is first checked against the local driving-rain index value. BS 5618, which is a code of practice for the use of urea-formaldehyde foam as a cavity fill, provides such design guidance and it is essential that the criteria stated in that Standard are followed. BS 5618 should be read in conjunction with BS 8208 Part 1 which covers those aspects that should be taken into account when assessing the general suitability of cavity walls for insulation. In some areas with a high driving-rain index, it may prove necessary to upgrade the external walls, for example by overcladding (Douglas 2006).

Alternatively, expanded polystyrene pellets and polyurethane foams are nowadays more commonly used to fill cavities (Douglas 2006). Prospective users of such materials would do well to use only those which have a British Board of Agrément (BBA) certificate and to check that the conditions of exposure to be encountered will be no worse than those to which the certificate relates.

When rain penetration does occur with urea-formaldehyde foam, it may be slight and, if so, it may be possible to remedy it locally by refilling the cavity at the appropriate spot. Otherwise, it will probably be necessary to increase the general resistance of the outer leaf to rain penetration by a suitable rendering or cladding. The expense of these latter remedies, however, may outweigh the financial advantages obtained by insulating: cavity-filling is an operation which it pays to get right the first time.

Mineral rock fibre treated with water repellent has also been used in regions with a high driving-rain index and, if properly injected, does not allow rain to penetrate. With all types of cavity fill it is essential that the outer skin of the structure is in sound condition otherwise spalling of the outer skin may occur after filling. It is also essential that cavities are filled by experienced and specialist tradesmen to minimise the risk of cavities becoming blocked by rubble falling from the face of the wall being drilled.

10.2.3   Built-in cavity insulation

The more stringent present-day requirements for thermal insulation can often be met conveniently by partially filling the cavity walls of buildings under construction with insulation boards or by wholly filling them with insulation batts. As might be anticipated, fewer problems are likely to occur with partial filling as there is at least some cavity remaining. The boards mainly used are of expanded polystyrene, but foamed plastics and glass fibre are also used. The boards are fixed to the cavity face of the inner leaf by clips attached to the wall ties or by nailing. If the number of clips or nails is inadequate, the boards will not lie firmly and flatly against the wall and will project into the cavity. The smaller the designed width of cavity the greater will be the risk that the projection will be sufficient to channel any drips from above onto the inner leaf. Of course, any such risks will be enhanced if the defects already mentioned in Section 10.2 are also present. When clips attached to the wall ties are used, four will be needed for each board, and wall tie spacings will need to match the dimensions of the boards. If boards are nailed, six nails per board will be the minimum needed. Again, prospective users would do well to use those boards which have a BBA certificate and to follow the restrictions which these certificates impose upon use. These are more stringent where the designed clear cavity is to be less than 50 mm. Then, if boards do not extend to the full height of the wall, cavity trays will be needed with weepholes above (see Section 10.2.3) to prevent rain from possibly dripping onto the top edge of the board and thence being channelled into the inner leaf. Insulation boards cannot be expected to lie flat against the wall if they are warped and this will happen if they are stored over battens; they must be stored flat.

Insulation batts used for total cavity fill are made from layers of glass fibres or mineral fibres treated with a water repellent. They have a somewhat laminated structure. There is, as yet, no great body of experience of rain penetration associated with their use but recent research has pointed to the likely hazards (BRE Digest 277). The main risk is when mortar used in the brickwork of the outer leaf extrudes into the cavity and adjacent to a joint between the batts. Such an extrusion will compress the batt as it is pushed in and will deform the laminations towards the inner leaf. Water entering through the outer leaf may drain down between laminations in the batts and follow these deformations to become closer to, and possibly reach, the inner leaf. It will be necessary to ensure that any such extrusions of mortar at the inner face of the outer leaf, but particularly those adjacent to horizontal joints between batts, are cleaned off as bricklaying proceeds. This will always be easier if the external brick face is taken up in advance of the inner leaf. If insulation batts are not taken up to the full height of the wall, cavity trays must be used with weepholes to protect the top edge of the batts. The batts themselves are not very robust and must be stored carefully under cover and not damaged.

10.2.4 Cavity trays at junctions

There have been numerous cases of rain penetration reported at junctions between walls and windows and doors, and where walls adjoin solid floors and ring beams. The former problem is considered in Chapter 9, which deals with external joinery. It is, however, appropriate to note here some difficulties which have been encountered when walls form a junction with concrete ring beams, columns and floors. Where cavity walls are built off concrete ring beams or floors, rain will usually penetrate the outer leaf and run down, and a cavity tray is needed to prevent the water from collecting on the beam or floor and feeding through the inner leaf, or along the floor, to the inside of the building. Cavity trays are also needed where the external leaf of a building becomes part of an internal wall at a lower level, a common occurrence when extensions are added to houses (see Douglas 2006) and in the construction of stepped terrace blocks. They are also needed at other points where the cavity is bridged, for example above airbricks and meter boxes. Many failures have occurred at these junctions for a variety of reasons. Sometimes cavity trays have not been provided at all which is a sure recipe for disaster. It should be noted here that house extensions and stepped terrace construction may be such that there is no room for a cavity tray of adequate depth to be provided. In such cases it is essential that the portion of wall which remains external is protected from rain penetration by other means. This may be by the roof above overhanging sufficiently or by cladding the external portion by a rendering or tile hanging. Other common causes of rain penetration have been

- the failure to provide a sufficient depth of upstand to the cavity tray;
- failure to prevent bridging of the cavity by mortar droppings;
- damage to the flexible materials mostly used, nowadays, in the construction of the tray;
- failure to support the tray by a mortar bed (haunching);
- inadequate lapping or sealing of lengths of cavity tray; and
- omission of stop-ends.

The traditional detail, with the tray sloping across the cavity, is prone to assist blockage of the cavity, and experience has indicated that a 75 mm upstand, previously assumed to be adequate, can be bridged by mortar droppings. Moreover, the traditional cavity tray was of metal and was supported in its diagonal passage across the cavity by a mortar haunching. Nowadays, it may be based on bitumen or polyethylene and the haunching is frequently omitted or poorly placed, causing the flexible tray to sag through lack of support. If mortar droppings are removed, there is a considerable risk of puncturing or distorting the tray, particularly where it is unsupported, thus making it ineffective. Rain has penetrated when lengths of the material used to form the tray have merely been overlapped and not properly sealed. Efficient sealing is, of course, difficult when the material is not supported. Penetration underneath the cavity tray is also a risk if this is not properly bedded, particularly if the surface of the concrete ring beam or floor is rough. Where cavity walls bear against columns, rain can penetrate at the junction between the cavity tray and the column face in the absence of stop-ends sealed to the latter (Figure 10.4).

Penetration from these causes is manifested by damp patches appearing after rain at or near to floor level and columns. The position of the patches may, or may not, however, be close to that of the original entry of the rain.

Failures may be prevented by designing cavity trays carefully in relation to the details of the construction involved. It should not be left to the man on site to "mock up" a solution: nowadays, this is not likely to result in success. Some general recommendations are

- that the upstand to the cavity tray should be a minimum of 150 mm;
- the tray should be dressed down the outer face of the inner leaf and then across the cavity – that is in an "L" shape for the tray rather than the traditional "Z" shape shown in Figure 10.1 (see Fig. 10.2);
- it should go over, not under, any flashings or DPCs associated with the detail (see Douglas 2006);
- the tray should be bedded in wet mortar; lengths should be sealed one over the other, not just lapped; and
- stop-ends should be provided at junctions with columns.

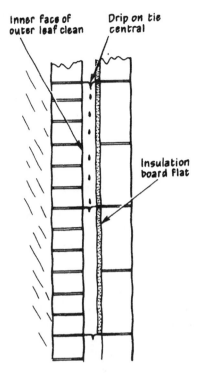

Figure 10.4 Good practice: tie centred; inner face of outer leaf clean; board flat and properly clipped. No rain penetration.

Duell and Lawson (1977) provide useful illustrations and the information on both the design and the installation of cavity trays and other forms of DPC.

Water penetrating along the external leaf and collecting above a DPC at the base of the wall, or above a cavity tray, needs to be drained to the outside. Weepholes have been the traditional method of drainage formed by omitting the mortar in appropriate perpend joints between bricks, usually every fourth joint directly above the DPC or cavity tray. Considerable amounts of water can collect in the absence of such weepholes and can feed through weak points in the system, particularly between unsealed laps in the DPC. Sometimes, weepholes have been provided but have been blocked by mortar droppings or cavity fill and made partially or wholly ineffective. If drainage cannot occur through weepholes, the water may drain out through the mortar joint above the DPC and it is probable that in many situations this can prove adequate. In exposed situations, however, some more positive drainage is needed. To prevent the possibility that rain will be blown into cavities through weepholes on very exposed sites, Duell and Lawson

(1977) refer to the use of tubes bent to prevent blow-back. They also expose the need to understand more fully the value of weepholes, but believe that it is safest to provide some form of drainage to cavities. This certainly seems desirable. The elimination of defects in the DPC system at the junctions mentioned is costly and most difficult, and greatly outweighs the extra effort needed initially to get design and construction right. Even the comparatively simple task of drilling out weepholes may cause damage to the cavity tray. It is recommended that pre-formed trays are used wherever possible and that all cavity trays are installed with stop-ends properly sealed to the tray.

10.3   Cold bridges and interstitial condensation

Bridging of a cavity wall by dense materials like concrete can result in cold areas and lead to condensation upon them. Areas particularly at risk include the surfaces of concrete lintels, ring beams and floor slabs which pass through to the external wall. The dampness will, in general, be more widespread in area than that caused by mortar droppings on ties, and will be localised at the level corresponding to the feature which bridges the cavity (BS EN 13788). It will, moreover, appear unassociated with rainfall. Such dampness needs to be avoided or minimised for the reasons given in Chapter 7. It may be possible to avoid cold bridging by designing so that some cavity is kept, or by providing extra thermal insulation on the inner side of the bridging feature (Figure 10.5). Such preventive measures are

Figure 10.5 Cavity tray sagging and unsupported.

more readily incorporated at the design stage than allowed for later, though it may be possible to protect dense lintels and ring beams by added thermal insulation (Douglas 2006).

The main points concerning interstitial condensation have already been made in Section 7.4. It only needs to be stated here that where vapour barriers are used they must be placed on the warm side of the insulation, and that wherever possible the permeability to water vapour of the external walls of a structure should increase towards the outside. Special care will be needed to ensure the continuity of vapour barriers at discontinuities in the structure.

10.4   Cracking and spalling of masonry through movement

Load-bearing masonry walls can crack and their surfaces may spall. Although such defects have not been amongst the most numerous, nationally, in recent years, there have been serious localised problems. There are reasons for believing that current construction methods and materials may lead to an increase in sensitivity to some of the agencies which give rise to these defects. Cracking of masonry is associated with movement and a principal cause is changes in moisture content of the units.

10.4.1   Movement due to moisture changes in the masonry units

Reversible and irreversible moisture movements in fired clay bricks are described in Chapter 6. Reversible dimensional changes are shown to be small – of the order of 0.02 per cent – and such changes do not cause problems in practice. It is the irreversible movement, which is usually many times as great, that has led to cracking. The extent of movement depends upon the type of clay, the degree of firing and the time which has elapsed since the bricks were removed from the kiln. As soon as bricks are removed and cooled, this expansion will start. Typical expansion in the first 2 days can range from 0.02 per cent for bricks made from London and Gault clays to 0.08 per cent for those made from Weald clays. Fletton bricks made from the lower Oxford clays have a typical 2-day expansion of 0.03 per cent. Such movement is around half the long-term, irreversible movement. As mentioned earlier, the corresponding expansion of brick walls as opposed to bricks is likely to be only one half of the latter (Figure 10.6). This should be taken as an approximate value, for the type of mortar and the restraint imposed upon the wall by the form of construction can also have an effect.

This irreversible moisture movement may cause several visible defects in brickwork. One typical result is an over-sailing of bituminous felt and polyethylene DPCs at the end of the wall, for such materials provide little

Figure 10.6 Cracking caused by moisture expansion of bricks.

restraint to expansion. This lack of restraint may also lead to cracking of walls near quoins. The sections of brickwork on either side of the return wall expand and tend to rotate the return. If this is short, for example less than 600 mm, vertical cracks develop, typically straight, and are visible above DPC only, but likely to extend to the height of the building. Cracking from the same cause and of similar appearance can occur near the corner of a building.

The majority of such problems can be overcome by not building clay bricks into a wall until at least 3 days have elapsed after withdrawal from the kiln and by avoiding the use of short return walls in long runs of brickwork. It should be noted that hosing down stacks of bricks and dipping individual bricks into water before laying is ineffective in preventing subsequent expansion. BS 5628 Part 3 calls for runs of clay bricks in walling to have a joint capable of accommodating 10 mm of movement about every 12 metres. As a general guide, the width of a joint in mm should be about 30 per cent more than the distance between joints in metres.

Where over-sailing of the DPC in the length of the wall has occurred, there is little that can be done but the movement is unlikely to be such as to cause instability. Cracking near quoins is also unlikely to be of structural significance and, when the cracks are fine, no repair is necessary. Where cracks are wide, it will usually be desirable to replace the cracked bricks. As irreversible expansion will be effectively over by then, the defect will not recur.

Calcium silicate and concrete bricks and blocks shrink upon drying and curing after manufacture and are also subject to reversible moisture movement, for which allowance needs to be made if defects are to be avoided. These defects take the form of vertical or diagonal cracks which commonly run from the corners of openings in walls, for example between upper and lower windows. If the mortar used is not too strong, the cracks will pass through the joints: otherwise, they may pass through the bricks or blocks, which is more disfiguring.

The presence of vertical cracks passing through bricks is a good indication that too strong a mortar has been used. Cracks appear early in the life of the building and their width is near to its maximum within a year. Calcium silicate bricks and concrete blocks have a high water absorption and many of the problems are caused by leaving them on site unprotected from the rain. They should be transported and stored under cover, be at least 2 weeks old and, also, be laid as dry as practicable. The tops of unfinished walling should be protected from rain. A mortar generally not stronger than a 1:2:9 cement:lime:sand mix is recommended unless frost is likely during construction, when a 1:1:6 cement:lime:sand mix or its equivalent in characteristics may be necessary. For calcium silicate brickwork, movement joints should be provided every 7.5 metres, though intervals of 9 metres may be satisfactory if the mortar has been chosen correctly. For concrete brickwork or blockwork a close spacing of movement joints to about 6 metres is preferable. Movement joints should be located at points in the structure where lateral support is provided. In parapet walls the spacing of movement joints should be half that of the main walling. In long runs of terrace housing, where the problem most commonly occurs, these will be at separating walls between the houses. Extra ties should be put in at 300 mm vertical spacing on each side of the joint, and the more flexible the wall tie the better, provided, of course, that the structural needs of the design are fully met. Cracking as a result of movement is repaired by either repointing the mortar joints or cutting out and replacing cracked bricks or blocks.

10.4.2 Movement due to temperature changes

Typical thermal movement of masonry walling is small, about 0.01 per cent for a change in temperature of 20 °C. Such movement does not cause defects in walls unless temperature changes are exceptional; therefore, it need not be considered further here.

10.4.3 Cracking due to roof movement

Cracking of walls caused by the spread of pitched roofs is not a common defect with new buildings. It occurs mainly through weakening of the structural members of a roof with time. Another cause, apart from general under-design, which is unlikely today, is the substitution of much heavier roof tiles for those originally specified. As the roof spreads, it moves the top few courses of the wall masonry outwards slightly; horizontal cracks appear in the plaster on the internal wall close to eaves level and cracking may also be noticeable externally. The main precaution necessary, nowadays, is to ensure that roof finishes are not changed without checking against the original design intentions. Should a defect from this cause appear, a detailed inspection of the state of the roof timbers, of the roof finish and the quality of the mortar joints will be needed before remedial work can be decided upon. It may well prove necessary to rebuild the roof and the top few courses of masonry. A more common defect associated with roof movement is that caused by the movement of flat concrete roofs. These can move appreciably with diurnal changes in temperature, particularly if the surface finish is dark in colour. The roof then tends to move outwards at the top of the walls and may push the top courses of masonry out slightly. Cracking then occurs within the internal plaster finish, usually near the corners of the building and close to the junction of the roof with the walls. Some cracking in the mortar joints of the external wall just beneath the roof may also be seen. It may be possible to prevent or minimise such defects by ensuring that any roof likely to exert a horizontal thrust on the wall is separated from it by a flexible DPC. When this is not feasible, the roof and the wall should be designed to act together. In either case, it will be advantageous to provide a solar reflective finish to the roof to minimise movement.

This defect is not of great significance, generally affecting only internal appearance. Where cracks are horizontal, and at the wall/roof junction, a coving may hide them. Decorating with lining paper and wallpaper may also prove successful to hide cracking which can occur lower down, probably near to the top of window openings.

10.5 Damage to walls by chemical attack

Most ordinary clay bricks contain sulphates of sodium, magnesium or calcium. These salts are soluble in water, calcium sulphate being less soluble than the other two. Normally, these sulphates are seen as the harmless efflorescence noted in Chapter 7, affect appearance only and need simply to be brushed away. Under wet conditions, however, the salts can react with tricalcium aluminate, which is always present in ordinary Portland cement, to form calcium sulphoaluminate.

Ordinary Portland cements vary in the amounts of tricalcium aluminate they contain, those with the highest amounts having the least resistance to attack by the soluble sulphates. The reaction is accompanied by a marked volume increase. Wet ordinary bricks, therefore, in contact with mortar based on ordinary Portland cement may lead to the formation of calcium sulphoaluminate and to disruptive expansion. It is, in fact, the mortar which is attacked, not the bricks, and the volume increase consequent upon the formation of calcium sulphoaluminate can cause vertical expansion of brickwork, which is commonly as high as 0.2 per cent. In theory, most brick walls with mortars based on ordinary Portland cement are liable to sulphate attack. In practice, fortunately, considerable sustained wetting is necessary. The most vulnerable walls are earth-retaining walls and parapet walls but sulphate attack is a problem also on rendered, and on facing, brickwork. The main sign of trouble on facing brickwork is cracking in the horizontal mortar joints, which generally occurs in a number of them. This may be preceded by some horizontal cracking in the plaster of plastered internal leaves of cavity walls, usually near eaves level. This internal cracking is caused by the expansion of the outer leaf in which the reaction is taking place, putting the inner leaf into tension. As the reaction proceeds, the external mortar joint spalls at the surface, and becomes weak and friable. The vertical expansion can result ultimately in spalling of the surface of the facing bricks and some bowing of the external walls. The brickwork may over-sail the DPC at the corners of the building, in a manner similar to that caused by the irreversible moisture expansion commented upon in Section 10.4.1. However, the latter movement will take place in the early months of the life of the building, while sulphate expansion is unlikely to appear for several years.

On rendered brickwork, sulphate attack is manifested by cracking of the rendering, the cracks being mainly horizontal and corresponding to the mortar joints below. The rendering may adhere quite well to the bricks early in the attack, but areas are likely to become detached as the expansion of the underlying brickwork causes severance of the bond between the two materials. There are three main ways of preventing sulphate attack in mortars: by ensuring that walls do not get, and stay, unduly wetted; by selecting bricks low in soluble sulphates; and by the use of cements low in tricalcium aluminate.

In most parts of the United Kingdom, the walls of structures between DPC and eaves level can be prevented from getting unduly wet by following established good design and construction procedures. These include the need to provide a good overhang to eaves and verges, to ensure that DPCs, flashings and weathering systems are effective and by keeping brickwork and blockwork as dry as feasible before, and during, construction. In areas where the driving-rain index is high (see Chapter 7), special precautions will be needed. In such areas, it will be desirable to give additional protection to the walls by rendering, tile hanging or similar treatment. Renderings generally keep the walls beneath them dry, and well-designed, and applied renderings will do so and help prevent sulphate attack. Those liable to shrinkage cracking, however, and dense renderings are particularly liable, can allow rain to penetrate the cracks but not escape readily afterwards. Under such conditions, the brickwork can remain wet for long periods, which assists the disruptive reaction to take place. The rendering itself may be attacked: expansion can then exceed 0.2 per cent. A 1:1:5–6 cement:lime:sand mix, or its equivalent, would be generally suitable for most conditions: denser, stronger mixes should not be used. Comprehensive recommendations for the design and execution of renderings on all common backgrounds are covered by BS 5262.

It is possible to obtain bricks low in soluble sulphates by specifying those designated as of "special quality" in BS 3921. Such a specification is likely to be considered too restrictive, however, for walling between DPC and eaves level, but may well be of value for brickwork severely exposed to rain, such as in chimneys and parapets. This is considered in Chapter 13, which is concerned principally with roofs, but in the context of which problems with chimneys and parapets are more conveniently covered.

Sulphate-resisting Portland cement and supersulphated cements can be obtained and, if these are used instead of ordinary Portland cement in the mortar, the likelihood of sulphate attack can be greatly reduced.

Where unrendered brickwork has expanded through sulphate attack, remedial measures will much depend upon the extent of damage. A first essential is to prevent the brickwork from continuing to become wet (and to eliminate the source of dampness, if this is a defect such as a leaking gutter or a defective downpipe). Generally, and in the absence of such obvious causes of dampness, it will be necessary to resort to tile hanging, ship-lap boarding or similar external treatment. When mortar has deteriorated considerably, repointing using mortars of composition 1:0.5:4.5 cement:lime:sand or 1:5–6 cement:sand with a plasticiser will be desirable. Treatment with a surface waterproofer is unlikely to be effective. Such treatment may, indeed, promote spalling of bricks containing large amounts of sulphates. They will, moreover, need frequent renewal.

When the rendering on masonry has deteriorated through sulphate attack, it, too, will continue to do so unless the source of dampness is eliminated and moisture is prevented from reaching the brickwork (Figure 10.7). When decay is localised, as might be the case if the cause were a leaking gutter, then some cutting-out and re-rendering may be feasible, once the cause has been dealt with adequately. Where sulphate attack is general, nothing will be gained by attempting to repair the rendering. It will be necessary to prevent rain from reaching the brickwork by tile hanging or some similar technique.

Brickwork is not affected by atmospheric pollution but stonework is, and the causes and defects are described in Chapter 6. When decay is superficial, it may be arrested by having the stonework cleaned by a specialist firm, which will use the most appropriate method in relation to the type of stone used, the extent of decay and the nature of the building. Such firms can also

Figure 10.7 Shrinkage cracking and sulphate attack on rendering.

restore decayed surfaces by plastic repair, but serious decay will necessitate removal and replacement of the whole, or a substantial part, of the stone block. It is generally not feasible to arrest the decay of stonework by the use of a stone preservative. Only for small areas of stonework of historical or architectural significance may it be possible, depending upon the particular types of salt contaminating the stone.

10.6 Damage to walls by physical attack

Water freezing within the pores of porous building materials, such as brick, stone, mortar and concrete, exerts a force, the magnitude of which much depends upon the amount of water present and the pore structure of the material. These expansive forces can result in defects but, as stated in Chapter 6, frost damage is comparatively rare between DPC and eaves level, though not uncommon in parapets. The winter of 1978–1979, however, produced more failures than usual, which may have been due to unusual combinations of sub-zero temperatures following prolonged rain. Under-fired bricks are more susceptible to frost damage than well-fired bricks and it is conceivable that over-enthusiastic attention to fuel economy may have contributed to the problem. In the general context of building failures, frost damage to walls has been of minor importance but the position needs to be watched with some care if winters become more severe, fuel economy remains a key issue and buildings become more highly insulated, for external walls may then reach lower temperatures more often and for longer periods.

Damage in buildings mainly occurs to brick, stone and mortar: concrete is seldom affected. Brick damage is seen usually as a crumbling or flaking of the surface and mortar suffers similar damage, which usually occurs before the mortar has had time to harden. Hardened mortar, though not immune if of incorrect strength, is not often attacked. Frost damage on limestone mostly causes small pieces to spall away but major cracking can occur. For most normal external walling in the United Kingdom, bricks of "ordinary" quality to BS 3921 should not be damaged, provided that, in winter work, they are not laid in a saturated condition. In severely exposed areas, local experience will provide the best guide. "Special" quality bricks to BS 3921, however, may be taken to be immune to attack in such areas. Special care in selection is necessary when bricks are imported into a new area, for example if English bricks are used in Scotland.

Local experience and advice from specialist agencies, such as the Building Research Establishment, are the best guides to the selection of frost-resistant stone. The problem is a minor one and, in general, affects limestones and not sandstones or the stones prepared from igneous rocks. The minimum quality for mortars for masonry is specified in BS 5628 Part 3. For external walls, a mortar consisting of a cement:sand mix of volume proportions 1:5–6, together with an air-entraining plasticiser, should not be damaged

Figure 10.8 Frost damage to bricks.

by frost. It should be noted that calcium chloride does not prevent frost damage to mortars and should not be used (Figure 10.8).

When damage has occurred, the most likely effect is to the appearance rather than to the stability of the wall. The individual bricks or stone blocks affected may be cut out and replaced, and mortar can be repointed, using a stronger and air-entrained mix. This is all that is usually necessary but, if many bricks are involved, it may be cheaper to cover them with tile-hanging or weather-boarding, or to render on lathing. Expansive damage can occur also when ferrous metals embedded in masonry, rust, and the effects on stone have been mentioned already in Chapter 6. Damage to stonework through corrosion can be prevented by ensuring that the metal fixings used are of copper, phosphor bronze, aluminium bronze or an appropriate austenitic stainless steel (see BS 5390). In brickwork, wall ties are used to provide structural interaction between the two leaves of a cavity wall and these are usually of galvanised steel. It has been stated already that zinc corrodes rapidly when embedded in black ash mortar. This affects, principally, that part of the tie in the outer leaf. Following loss of the zinc, the mild steel corrodes, and ties of the vertical-twist type contain sufficient steel for the formation of the rust to cause the external leaf of the brickwork to expand vertically (Figure 10.9). The total expansion in a two-storey wall may be 50 mm. This

Figure 10.9 Corroded wall ties.

results in horizontal cracking of the mortar joints in the outer leaf, coinciding with the level of the ties, usually four courses apart, and with cracks several millimetres wide. This cracking contrasts with that due to sulphate attack by being at these regular course intervals. There are usually no vertical cracks. In severe cases, the inner leaf may also show cracking and structural integrity can be impaired. Calcium chloride in cement/lime/sand mortars can also promote premature corrosion of ties. It is better not to use aggressive mortars but, where this cannot be guaranteed, it is advisable to use non-ferrous ties or those of stainless steel. However, even in normal cement mortar, corrosion can occur and cause cracking in walls which are severely exposed and if sub-standard ties are used. All metal ties used should conform to BS 1243.

The action necessary following observation of the defect may vary from doing nothing to complete rebuilding of the outer leaf and replacement of all the old ties. Much will depend on whether the damage affects mainly appearance or stability. Intermediate courses of action include piecemeal replacement of corroded ties by more resistant ones and the application of cladding to reduce the wetting of the wall by rain. A proprietary foamed plastic injected into the cavity wall is also claimed to be a remedy, as well as providing some valuable thermal insulation, but long-term experience of its behaviour is at present lacking.

10.7 Problems with renderings

Sulphate attack and its effects on renderings are considered in Section 10.6. Renderings may crack, craze and become detached from their background, however, even in the absence of sulphate attack, through differential movements between the background and the rendering. Cement-based renderings shrink as they dry, particularly the strong, dense mixes such as a 1:0.25:3 cement:lime:sand mix. The effect of this shrinkage of renderings on rigid backgrounds is to set up stresses which may be relieved by cracking or by loss of adhesion. The cracks may then allow rainwater to pass through them but prevent its ready evaporation afterwards, which can assist sulphate attack, as already described. Cracking from shrinkage, and in the absence of sulphate attack, is likely to be random and, if tapped, the rendering may sound hollow (Figure 10.10). Differences in drying shrinkage characteristics between the top coat and the undercoat of renderings can also lead to defects. If the top coat is richer in cement and, thereby, stronger, with a greater drying shrinkage, it can pull away from the undercoat over small or large areas. When the undercoat is considerably weaker, parts of it may come away with the top coat.

Failures can be avoided by carefully following the recommendations made in BS 5262. For most situations, mixes equivalent to a 1:1:5–6 and 1:2:8–9 cement:lime:sand will be satisfactory, if it is ensured that the final coat is not stronger than the undercoat. Detailed recommendations are given in Tables 1 and 2 of BS 5262, which specify suitable mixes for renderings in relation to different backgrounds, exposure conditions and the texture of finish required.

Remedial work, when cracking does not extend into the background, necessitates the cutting out of hollow areas adjacent to the cracks and patching with a suitable rendering mix. Where cracking extends into the background, the latter will first need making good, or it may be possible to fix a waterproof lathing to the background and then to render. Flaking due to the use of too strong a finishing coat is likely to need the complete removal of the finish, together with any adherent undercoat, and its replacement by a mix no stronger than the undercoat. The surface of the latter may need to be roughened and its suction reduced by the prior application of a spatterdash mix.

When corner beads are used to provide good edges and corners for a rendering, these should be of a material suitable for external exposure, preferably stainless steel. Corrosion, followed by spalling, can occur if beads only suitable for internal plastering are used. Failed renderings are not generally very expensive to put right but, unfortunately, this is far from being the case for failures of the many other forms of cladding considered in Chapter 11.

Figure 10.10 Random cracking of rendering through excessive drying shrinkage.

10.8 Summary

Walling is susceptible to attack by atmospheric and mechanical agencies. Wind-driven rain, sunshine or frost action can easily degrade wall surfaces and can even cause problems below the substrate. Impact damage is another high risk factor for walls. Any such defects are usually conspicuous even if their cause is not.

Chapter 11

Cladding

OVERVIEW

This chapter deals with cladding problems. Like the roof it is a conspicuous part of a building's exterior and is highly exposed to many of the degradation mechanisms highlighted earlier.

11.1 Background

Scarcely a week goes by without accounts in the technical press, and in the columns of local newspapers, of expensive repairs needed to cladding on high-rise buildings. It is not uncommon for costs exceeding £1 million to be quoted as the sum needed to rectify faults on one high-rise building: the total national cost must be very high.

Many such buildings are now surrounded by scaffolding and safety netting to prevent pieces of cladding, thought likely to break off, from falling to the ground, possibly injuring pedestrians. Even a kilo or so of brick or concrete falling from 10 storeys may cause more than alarm and despondency – and some panels weigh well over a tonne.

Cladding defects are caused by differential movement between the cladding and its background; by failure to allow for the inaccuracies inherent in construction; by inadequacy of the fixing and jointing methods used; and by premature failure of sealants.

11.2 Differential movement

11.2.1 Background

There are several causes of movement in claddings and of the backgrounds to which they are fixed. It is the relative movement between the two which is of first importance. The fact that substantial relative movement can occur in a tall building and needs to be accommodated has, unfortunately, been overlooked in the design of many post-war buildings. The principal causes of movement are due to temperature effects, moisture changes and creep (Bonshor and Bonshor 1996). Failures have been mainly in cladding over reinforced concrete structures, most probably because steel structures have no moisture movement and creep is small.

11.2.2 Temperature effects

The thermal movement of common building materials is shown in Table 5.4. Cladding, being more exposed to the weather, is likely to be subjected to greater movement, either of contraction or of expansion, than the structural backing and, particularly so, when the cladding is thin and the structure massive. The movement may take place rapidly or may be more seasonal in character and, as shown in Chapter 5, will be dependent not only upon the inherent nature of the material but also upon its colour and the extent to which it is insulated. For example, significant differential thermal movement can occur and cause distortion when stones of similar coefficients of thermal expansion, but of different colour, are used. Such problems have been reported when dark marble and light granite have been used in alternate strakes, and with travertines and slates of different colours. Most cladding will be subject to a temperature range in service of 70 °C in the United Kingdom, but if highly insulated on the inside, the range can exceed 100 °C. Cladding is restrained to some extent from taking up the change in size consequent upon the change in temperature. This restraint causes stresses which, if sufficiently great, may cause distortion or fracture. The extent and nature of the damage caused will be dependent upon the physical and mechanical properties of the cladding, and the extent of the temperature change. It may also be affected by the time over which the temperature change occurs and the frequency (Figure 11.1).

11.2.3 Moisture effects

Cladding is fixed, in general, to steel or concrete structures. The former has no moisture movement but concrete shrinks after placing, as its moisture content slowly reaches equilibrium with the surrounding atmosphere. An irreversible shrinkage of between 0.03 and 0.04 per cent can be expected (BRE Digest 228) for normal gravel aggregate concretes. Exact values for concrete depend upon the composition of the concrete mix and the particular aggregates used. Concrete also possesses a reversible moisture movement of between 0.02 and 0.06 per cent but the concrete structure is unlikely to become wet in use, for it is protected by the cladding. It is the irreversible drying shrinkage of the structure which is of principal importance, and slow shrinkage will occur over a long period. The moisture movement of common cladding materials is not usually the cause of problems. The

Figure 11.1 A fine sense of timing–falling cladding.

irreversible expansion of freshly fired bricks already described can, however, be a contributory cause of failure.

11.2.4 Creep

It is probable that a major cause of cladding defects has been the failure to remember that structural concrete creeps under the dead and imposed loads placed upon it; that is, a sustained load produces a permanent deformation. This takes place over a long period of time, though the bulk will occur during the first 5 years. Creep is a complex phenomenon and the extent of the deformation is related not only to the applied stress in the concrete but also to the time after placing the concrete, the nature of the concrete mix, and the type and placing of any reinforcement. Within the normal design stresses for concrete of 20–$35\,N/mm^2$ at 28 days, the average value of creep can be taken as $30 \times 10^{-6}\,mm/N/mm^2$ (BRE Digest 228). Creep of concrete at, say, $30\,N/mm^2$ may result, therefore, in a deformation of 0.09 per cent which, added to the irreversible drying shrinkage, can give a total shrinkage of around 0.12–0.17 per cent for the structural concrete frame. Thus, the shrinkage in a 3-m-storey height could be as much as 5 mm, and 50 mm in a high-rise block of ten such storeys. Additionally, there can be a small elastic deformation of the concrete structure under load (mainly affecting beams rather than columns).

11.2.5   Effects of differential movement

The effects of a combination of drying shrinkage and creep in the structure, and of expansion in the cladding, may be illustrated by the case of brickwork cladding to reinforced concrete. Failures in such a combination have, in practice, been all too common in recent years. The coefficient of thermal expansion of brickwork vertically is likely to be around 7×10^{-6} per degree C and the temperature range likely to be encountered, in service, with fired clay brickwork cladding might reasonably be taken as 70 °C. The maximum expansion of the cladding will depend, inter alia, upon the temperature at which it was first placed: a maximum vertical thermal expansion of 0.05 per cent is possible. This will act in opposition to the shrinkage of the concrete structure to which it is attached. In total, a differential movement of some 0.22 per cent could occur – that is, around 6 mm in a storey height, from these causes. Brickwork may also exhibit an irreversible moisture expansion, as already described, and the shrinkage of concrete can be greater if shrinkable aggregates are used.

The frame of a new timber-frame house will shrink as the moisture content of the timber reduces to the level reached when the house is in full occupation. This shrinkage will be enhanced if the frame is not kept protected and dry on site. In a masonry-clad timber-frame building, differential movement will occur. Distortion and subsequent rain penetration will then take place if gaps and joints are not properly designed and executed between the masonry and the frame. Places particularly at risk are at eaves, sills, jambs and heads of openings. Flexible wall ties will need to be used between the frame and the masonry.

11.3   Inaccuracies in construction

The likelihood of failure has also been enhanced by not allowing for the dimensional deviations which occur in manufactured or site-cast products, and for the inaccuracies in setting-out and during erection. In the past, corrective measures to obtain good fit were taken on site almost as a matter of course, but, with modern multi-storey buildings, this is less possible, mainly because of the greater inflexibility to manipulation of the products used.

Supporting features, compression joints and fixings for cladding have often not been designed or placed to allow for the likely inaccuracies over the whole building, and these can be additional to the differential movement already described. A particular example is that of insufficient bearing on the floor of masonry cladding, designed to oversail the floor. Cladding is used in this way, together with fired clay "slips" to hide the floor edges, when the design requires the elevation to show as a continuous masonry face, masking the structural framework behind. Excessive stress has also been caused to supporting nibs when the heavier cladding units of stone and precast concrete have had an inadequate bearing on them. Reinforcement is often not taken out far enough into the projecting nib, which has led to sheer failures. Although the faces of concrete nibs are shown generally as being square in design drawings, this is often far from the case in reality. Mortar used to dub out the nib to the correct profile will not be reinforced and does not provide an adequate bearing to the cladding units. Inaccuracies in construction have also led to the displacement of fixing devices and failure to attach cladding adequately to the supporting structure. Many fixing devices can be adjusted only within comparatively narrow limits.

11.4   Faults caused by movement and inaccuracy

Many failures have occurred because the design and construction of the structure, its cladding and the fixings have been insufficient to allow for the likely total movement. Sometimes, mistakes during construction, such as forgetting to remove packing spacers after final fixing, have also contributed. As a consequence, a compressive force has been transmitted to the cladding which has not been designed to accept it. Characteristic failures, due to this squeezing effect, on brickwork used as cladding to reinforced concrete structures show as horizontal cracks at roughly storey-height intervals on the line of the floor slabs, though all storeys may not be affected. This cracking is frequently accompanied by spalling of the edges of individual bricks in that vicinity, and buckling and displacement of a few courses of brickwork (Figure 11.2). Brickwork may be supported on concrete nibs projecting as part of the reinforced concrete floor slab and, although uncommon, excessive relative movement can cause cracking of the nibs, especially if they are weakened by corrosion of any reinforcement present. Where

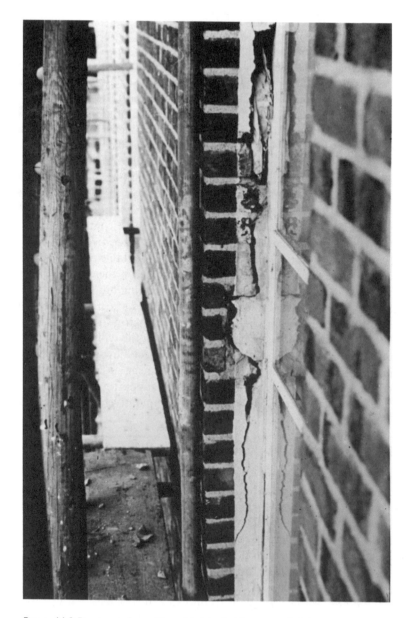

Figure 11.2 Bowing and cracking of brick cladding to an RC frame.

brick "slips" have been used to cover the edge of the floor slab, these often buckle badly and spall and may, indeed, justify their name by becoming completely dislodged (Figure 11.3). Flexible DPCs used at floor level are often extruded. The defect is one which is unlikely to appear before

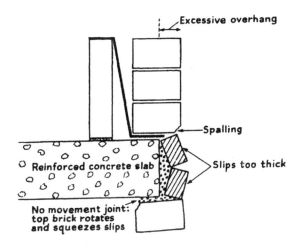

Figure 11.3 Buckling of brick slips. Creep and shrinkage of concrete frame and expansion of brickwork causes spalling of bricks and buckling of slips if adequate compression joints are not provided.

several years have elapsed after construction. Similar effects can appear when natural stone or reinforced concrete is used for cladding to reinforced concrete structures and when mixed cladding materials are used. In one case reported (Department of the Environment 1973), the compressive stresses generated in brickwork cladding, assisted by inaccuracies in construction, caused bending forces to be applied to an unreinforced precast stone band course, causing major cracking and failure. A further cause of failure with reinforced concrete cladding panels has been due to corrosion of the reinforcement in the panels. This has been assisted by inadequate general cover to the steel, particularly where exposed aggregate of a large size has been used as the finish. In addition, compression of the edges of the panels, through the differential movement mentioned, has caused cracking and spalling and thus a reduction in the cover, leading in turn to further corrosion.

Mosaic cladding sheets may also fail from the same cause and tend to crack rather than to spall. As the adhesion between the mosaic sheets and the background is usually not strong, bulging and total detachment are the predominant defects. Ceramic tiles are commonly used for external cladding. The mode of failure due to differential movement depends to a large extent upon the type of tile. Pressed tiles lack the good undercut rear face which most extruded tiles possess and tend to become detached at the interface between the tile and the bedding mortar. The more commonly used extruded tiles, together with adhering mortar, tend to come away from the structural background. Failure of brick slips, mosaic sheets and ceramic tiles is also likely if fixing and adhesion are poor (see Section 11.6.2).

Glass-reinforced cement (grc) cladding panels are made from alkali-resistant glass fibres, usually in the range of 34–38 mm long, cement and sand. A field survey of grc cladding panels in use in the United Kingdom showed an unacceptably high proportion of buildings inspected to have some cracked panels (Moore 1984). In part, this was due to not only some deficiencies in manufacture but also undue restraint imposed by fixings and a failure to recognise the effects of adverse combinations of temperature and moisture change.

Glass used as lightweight cladding in curtain walls can crack through differences in the rate of response to changes in temperature between it and the framework of metal into which it fits. Metals have a higher coefficient of thermal expansion than glass, often twice as much, and may exert sufficient force on the glass to cause it to crack, if clearances between the glass and the frame are too small. The crack is usually a single crack, starting from the edge of the frame.

11.5 Sealants

Flexible joints used in cladding are affected both by movement and by inaccuracy. The different sealants used for jointing have properties which allow certain movement to be accommodated safely in relation to the width of sealant used. This movement, expressed as a percentage of the minimum width of sealant necessary, is shown in manufacturers' literature as a "movement accommodation factor". The wide variety of sealants available covers a wide range of movement accommodation factors but, taking a fairly typical factor of 20 per cent, the minimum width of sealant necessary in the joint would then need to be five times the likely total movement. There have been many failures of sealants through failure either to determine correctly the likely total differential movement or, having determined it correctly, to use a width of joint which can accommodate that movement but is, nevertheless, too narrow for the sealant chosen. A sealant unduly stressed will tend to break down and its durability will suffer. This, in turn, may enable rain to penetrate, which would not otherwise happen.

11.6 Fixing methods

11.6.1 Background

The weak point in any cladding system is the fixing. Cladding is fixed to the structure, usually by metallic fasteners of various kinds or by adhesive.

11.6.2 Metallic fasteners

The main defects associated with metallic fasteners have been their inability to perform the design function satisfactorily and their failure through corrosion. The inaccuracies inherent in the construction of high-rise buildings require that fixings have a fairly wide range of adjustability but many do not, and the measures taken on site to get a fit are often unsatisfactory. Common faults are the lack of full and proper engagement of dowels in the slots or holes meant to receive them; angled surfaces instead of square ones, which makes good bolting difficult to achieve; and too great a thickness of packing pieces used on bolts, which can reduce the efficiency of the latter. Sometimes, holes or slots incorporated in the fixing device, compressible washers and long-threaded bolts, all used to allow adjustment to be made, cannot be properly utilised because of constraints in adjacent parts of the construction. Typical problems have been illustrated by Bonshor (1977). More in the realm of carelessness than inaccuracy is the failure to tighten bolts properly or sometimes to overtighten them which, in the latter case, can cause excessive stress concentration on the cladding material.

Corrosion of metals is considered in Chapter 6. Cladding panels are used in exposed conditions and those based on heavy, porous materials, such as stone and concrete, tend to be on the thin side to keep down the overall weight. It may be expected that, under those conditions, the fixings used with them will be wet for long periods and in an environment conducive to corrosion.

Corrosion may be caused by well-known effects, including electrolytic attack which may occur if packing pieces of one metal are left behind to react with different metals used for the main fixings. The combined effects of corrosion and stress can cause stress corrosion cracking. Atmospheric pollution and moisture, coupled with stress corrosion, have been known to attack manganese bronze cramps used with stonework. Corrosion of fixings causes staining on the face of the cladding and the cracking or spalling of brick, stone or concrete cladding close to the points of fixing. It thus gives useful prior warning that a thorough investigation of the state of fixings is necessary to avoid complete detachment of panels. However, excessive force on non-ferrous bolts, due to lack of adequate allowance for movement of cladding slabs, has led to shearing of the bolts and sudden detachment of the slabs without such warning.

11.6.3 Adhesives

Brick slips, mosaic sheets and ceramic tiles are the commonest forms of cladding attached by mortars or organic adhesives rather than by mechanical fasteners. The differential movement of the structure and the cladding due to thermal, moisture and creep effects are described above. The shear stresses imposed on brick slips, mosaic sheets and ceramic tiles have often destroyed the bond between them and the mortar or other adhesive used, or between the latter and the structural background. It is probable that water reaching bedding mortars and freezing, in the severe exposure conditions to which claddings are often subjected, has also contributed to failures of bond with brick slips which are porous and allow water to penetrate. Inaccuracies in the nib of the floor slab to which the slip, mosaic or tile is to be stuck can lead to the need for either excessively thin or thick adhesive beds, neither of which contributes to success. Thicknesses required for bedding mortar for brick slips to keep them in the same vertical plane as the rest of the cladding have been found to range from 3 mm to more than 25 mm. The preparation of the background is of critical importance. Failures on concrete backgrounds have been assisted by the preparation and cleaning being insufficiently thorough in removing mould, oil and laitance. Dense structural backgrounds, too, have low suction, which has led to difficulties in the adhesion of mortar beds. Unfamiliarity with the many different organic adhesives available has also contributed to failure. Contributory causes have included the incorrect choice of adhesive; storage which has been too long before use or at temperatures which cause deterioration; and lack of control in the proportioning of the various ingredients which go to form the adhesive. Many organic adhesives have two ingredients which have to be mixed accurately, and in the right order, before use. Failure to do so has led to lack of early strength and a short life in service. Proper application of the adhesive is required: often too thin a coat has been used or the adhesive has been dabbed on at a few points rather than spread completely and evenly over the cladding unit. For brick slips to floor nibs or edge beams, the three adhesives mainly used have been based on epoxy resins, polyester resins or sand/cement mortars modified with styrene/butadiene rubber. Epoxy resin adhesives are two-part, or sometimes three-part, adhesives which need to be carefully proportioned, mixed in the right order and, to be effective, applied to a thoroughly clean but roughened surface: support is also needed for up to 24 hours. These are not properties to which normal site working and workmanship are sympathetic. Polyester resin systems vary considerably in their properties, particularly in their working life after mixing and in their ability to bond successfully to damp surfaces. The ratio of resin to hardener is often critical. The sand/cement/styrene/butadiene rubber adhesives are closer to normal mortar mixes in their preparation and application and can be used on damp surfaces. Their strength development is low at low temperatures. Failures using these adhesives have been caused by too thick an adhesive bed due to inaccuracies in the positioning of the concrete floor nib; too smooth a face to the floor nib; lack of adequate grouting of the brick slip and of the nib; and inadequate coverage of the slip by the adhesive. Delay in positioning the slips after applying the adhesive can also lead to detachment. Failure from this delay may be assumed when the pattern of the adhesive formed by using the customary notched trowel can be seen to be still undisturbed on the failed slip.

Mosaic sheets come in two main forms: paper-faced mosaic in which the pieces of mosaic are glued face down to paper, or bedding-side down to nylon strips or nylon fabric (nylon-backed mosaics). In the latter case, the backing is embedded in the mortar or adhesive. Failures occur from the same general causes as with brick slips. Additionally, if final straightening of the joints is done after the bedding has started to set, the latter can be stressed and the bond broken between it and the mosaic. It is a mistake to suppose that, because the face of the mosaic is relatively impervious, rain will not reach the back. The large number of joints between tesserae allows rain to penetrate and pass into the building mortar or adhesive, from which position it can be slow to escape. This will enhance the possibility of frost and sulphate attack if mosaics are applied to brickwork containing soluble sulphates. There have been spectacular and costly failures of mosaics from the latter cause.

11.7 Prevention of loss of integrity in cladding

Ways of reducing the likelihood of failure are fairly self-evident from the foregoing sections. The first essential is to be aware of, and to calculate, the extreme range of differential movement likely to occur between cladding and background, and to ensure that the compression joints and movement joints required can cater for this, as well as for the inaccuracies likely in manufacture and construction. Guidance on the latter is available in BS 5606. The width of joints will, in addition, need to be sufficient to allow movement so as not to cause undue stress to the sealant to be used, and reference will need to be made to manufacturers' information, in particular, to the movement accommodation factors. Compression joints will also need to be wide enough and filled with materials which do not compress so much that unwanted forces are transmitted to the cladding. Minimum compression-joint thickness needed per storey height is likely to be around 12–15 mm for brick or precast concrete cladding applied to a reinforced concrete background, but these values are guides only and should be considered as probable minima in normal construction. When the cladding is to be supported directly by the structural frame, it is essential to provide a bearing of at least 50 mm after allowing for all the dimensional deviations probable. The structural consequences of the failure or omission of one or more fixings should be evaluated. Where precast reinforced concrete cladding is used, cover to reinforcement should be a minimum of 25 mm and, generally, in accordance with Table 1 of BS CP 116. Where exposed aggregate finish has been used containing stones of 75 mm or over in size, the reinforcement should never be less than 20 mm behind the stones. Methods adopted for fixing claddings to backgrounds are many and, often, complex. It is essential that, at design stage, careful thought is given to the inaccuracies likely to arise in construction and the effect these will have on the ability of the fastening device to hold the cladding securely. These inaccuracies are likely to be greater than the designer anticipates and he should consciously increase his own estimate of them. Fastening devices should be chosen which have a high adjustability and, in design, provision for adjustability should be considered separately from provision for movement during service. The aim should be for an adjustability which can be made easily on site and for the integrity of fixings to be capable of being checked before being covered by the cladding. The fewer the fixings necessary the better. Portions of fixings in the structural frame need to be positioned with great care, which implies a high degree of site control. The critical alignments and dimensions of fixings, and of sockets in particular, need to be specially identified and quantified in design, and to be made known to those involved in the construction process. No changes should be improvised on site without prior reference to the designer. Useful guidance on fixing problems is given in BRE Digests 223 and 235 and by Bonshor and Eldridge (1974) in their work on tolerances and fits.

Designers should assume that fasteners in contact with cladding will get wet and, consequently, should be chosen to remain free from corrosion under these circumstances. The possibility of bimetallic corrosion, particularly if there is any mild steel involved in the construction, should be evaluated. Metals shown in BS 8297 should be suitable for fasteners. They include phosphor bronze, silicon aluminium bronze, copper and certain types of stainless steel. The use of ferrous metals is permitted in that Code but with a qualification which seems to be of doubtful real value. It would seem better to avoid the use of ferrous metals altogether in view of the danger if fixing devices fail, and the high cost of remedial works. It would also be desirable for the design to incorporate ways by which the condition of any fixings can be inspected during their lifetime without undue disruption and difficulty.

The best advice on brick slips is not to use them but, if aesthetics overrules prudence, then the designer and client should understand that the adhesives used so far cannot be guaranteed to be successful. Much will depend on the specific design, but even more on the site workmanship and control, the choice of adhesive and its proper mixing and use. Failure sooner or later is more likely than success. Success may be more probable, however, if some form of mechanical anchor is used instead of, or in addition to, adhesives, though these may spoil the unbroken line of the slips. To have much chance of success with mosaic and ceramic tiling the detailed recommendations given in BS 5385 need to be followed carefully, as do the precise recommendations of the manufacturer of the adhesive when organic adhesives are to be used. These will usually cover such items as the type of trowel to be used, the mixing procedure, the working time after the adhesive is spread and the suitability of the background. As with brick slips, failure is, nevertheless, more probable than success and there are more constraints to using mechanical support systems. To minimise the risk of cracking of grc the coefficient of thermal expansion should be taken as $20 \times 10^{-6}/°C$, the reversible moisture movement around 0–15 per cent and the irreversible drying shrinkage 0.05 per cent. There will be less risk of cracking if small, flat sheets are used than larger panels of more complex section. Where glass is used for cladding, the most common cause of failure can be prevented by ensuring that clear glass has a clearance all round of at least 3 mm. Coloured glasses, or clear glass with a dark background close by, can reach temperatures as high as 90°C even in the United Kingdom. Greater movement must be allowed for – not less than 5 mm where the longer dimension of the glass exceeds 750 mm.

It is not useful to generalise on the remedial measures necessary once cladding has failed. Many variations of systems of cladding and structural frames are possible, and failures have ranged from those where complete replacement has been necessary to those where local patching has sufficed. It is true to say, however, that most cladding failures are highly expensive to rectify. The need to estimate properly the likely differential movement and inaccuracies in construction, and to allow for them adequately, demands special emphasis (Brookes and Meijis 2006). So does the need to ensure that reinforced concrete cladding panels have wholly adequate cover to reinforcement, particularly if an exposed aggregate finish is used.

11.8 Water entry

Many cladding systems have proved incapable of preventing the entry of rain (Endean 1995). In part, rain penetration has followed the cracking, spalling and dislodgement of cladding caused mostly by the differential movement described. Lack of attention to joints and to rain-shedding features has, however, also been a cause. There are many forms of joint associated with the various claddings used, some complex in geometry (which adds to the likelihood of difficulties in site work). Common problems of rain penetration, however, may be illustrated by reference to three main classes of joint – the butt joint, the lap joint and the drained open joint. The butt joint, formed when units are simply butted together, is the simplest and relies on a gap-filling material, usually a mortar, organic sealant or gasket, to permit movement and to keep out the rain. The gap-filling material has to take the full movement of the joint. Cement-based mortars shrink and their adhesion to most backgrounds is not particularly good: generally, one would not look for their use with large cladding units. Failures with organic sealants have been due mainly to their use in butt joints which are too narrow and too deep, leading to undue stressing, as described in Section 11.5; this, in turn, has led to breakdown. Inaccuracies in panel manufacture, particularly in joint width, and in fixing on site which has led to panels being out of true, have contributed to failure. Inadequate cleanliness and inappropriate surface texture of the cladding have led to loss of adhesion and allowed leakage.

When cladding panels lap across one another, the lap joint formed imposes less strain on the sealant and this is also partially protected from the weather. Failures have been due largely to lack of good adhesion of the sealant to the surfaces of the cladding units. Gaskets are pre-formed materials based on rubber or plastics, and depend not upon adhesion but on being squeezed by the units to provide the seal. They are used mainly with large cladding units, particularly pre-cast concrete panels, in drained open joints but may also be used in lap joints. Gaskets cannot adjust well for irregularities in joint width, or for excessive roughness of the surfaces to be joined. The joint at the intersection of vertical and horizontal gaskets can be difficult to design and achieve on site. The drained vertical joint consists generally of an outer zone, essentially to trap rainwater and provide drainage, and an inner zone which provides an air-tight barrier. This also is intended to prevent entry of any rain which has penetrated past the outer zone and the baffle which is commonly provided (Figure 11.4). The materials used for baffles need to have a long life, for they are usually difficult to replace. The air-tight barrier is formed by a sealant or gasket and is not exposed to weathering. This inner barrier is essential to the success of the joint and needs to be air and weather-tight to prevent any water which reaches it from passing by capillarity into the building. Drained open joints have flashings provided to direct to the outside any rain which has

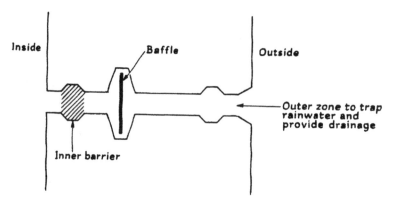

Figure 11.4 Plan of a typical drained vertical joint. The baffle is often difficult to replace.

passed the baffle. Horizontal joints are usually protected by an upstand, so that rain cannot blow directly through the joint. This, generally, takes the form of the lower edge of the upper unit projecting downwards to some 50 mm below the top of the upstand of the unit below. Wind and rain conditions around high-rise buildings, in which large-panel construction using drained open joints is mostly used, are severe, and penetration of rain has occurred through failure to make the internal air seal effectively, because of difficulty in access to the back of the joint; by failure of the flashing to bring the rainwater to the front of the joint; by omission of the flashing and inadequate lapping and sealing; by inadequate depth to upstands; and by baffles at the intersection of vertical and horizontal joints failing to reach down far enough, thereby reducing the effectiveness of the upstand. Many of the difficulties have occurred through failure to recognise the tolerances inherent in manufacture and the inaccuracies in construction.

Weather-strips or throatings are sometimes used as horizontal projections, their purpose being to shed rainwater clear of the units below. Failures have happened when they have collected rainwater which has been blown by high winds through the joints. When water does appear at the inside of a building, the path of its entry may not be obvious, and a detailed and costly examination is often necessary to identify the weak points in the cladding system. Water appearing at a low level may well have drained from above or, in some forms of curtain walling, have travelled considerable distances through hollow sections. In curtain walling, wind pressure on glazing can cause displacement of glazing compounds when distance pieces are omitted or too narrow. This has allowed rainwater to be pumped past the compound, especially at high levels. Distance pieces are commonly made of plasticised PVC and are resilient. They need to fit tightly between glass and frame. Design recommendations are given in BS 6262.

Water penetrating through cladding can be difficult to trace and expensive to rectify (Endean 1995). With heavy cladding panels, it may, indeed, never be possible to gain access to, and replace, sealants, gaskets, baffles and flashings. Problems arise, in the main, at junctions of cladding units with one another and at openings. Design detailing here needs much care and should not be left to inexperienced designers. It must take into account tolerances and site inaccuracies and be as simple, geometrically, as possible. Design should strive to allow reasonable access to key parts of the water-barrier system, for even the most durable sealants, correctly used in the proper joints, are likely to have a life expectancy less than that of the building of which they are a part. Close liaison between designer and contractor on the feasibility of joint construction in relation to the proposed design is vital. Jointing is not something which can be left to the site operative to fudge as best he can.

11.9   Metal cladding

11.9.1   Surface degradation

As reported by Noy and Douglas (2005), since the 1970s, profile metal cladding has been used extensively for roofs and wall panels in many commercial and industrial buildings. The main protection and colour medium of this type of cladding is usually a plastic coating, either PVC (e.g. "Plastisol") or PVDF (polyvinylidene-fluoride). A typical initial service life for this form of cladding is between 15 and 25 years, before first maintenance is usually needed (Harrison 1996).

The main surface degradation problems are

- Delamination of sheeting and insulation
- Denting or puncturing of sheeting
- Discolouration or bleaching of surface coating
- Peeling and blistering of surface coating
- Pitting of the surface
- Rust staining – particularly at fixing nodes and at the cut edges.

11.9.2   Cut edge corrosion

Apart from surface degradation, one of the most common problems in metal profile sheeting is cut edge corrosion (Noy and Douglas 2005). It occurs where sheets are cut to length in the factory or on site, thus exposing the unprotected metal substrate. This gives rise to the telltale rusting along the joints of the sheeting (Weatherproofing Advisors Ltd 2004). Not only is this unsightly, it will also lead to further degradation of the sheeting and substrate causing eventually leakage at the joints.

Increased environmental pollution from acid rain, etc. dissolving the protective zinc oxide layer of the galvanised substrate is the main deterioration mechanism that triggers this problem. This can be exacerbated by the use of less durable organic coatings such as "Plastisol" or poor workmanship when installing the sheeting (Harrison 1996).

The three main areas of cut edge corrosion attack on a roof detail are

- Overlaps (including those above translucent sheets)
- Gutter edge detail
- Flashing detail (especially those edges not turned over).

There are now a number of proprietary treatments of edge corrosion available (for details, see Noy and Douglas 2005). Delvemade Ltd is one such system (www.delvemade.co.uk), which uses a silicone sealer application. The sealant must be of a non-acidic type to prevent it from triggering corrosion again in the treated edges of the cladding. This repair technique can be used in conjunction with a recoating scheme for the whole external surface of the cladding using a more durable paint system such as PVDF.

Consult also the technical guides by Weatherproofing Advisers Ltd (see details in Bibliography) for feedback about remedial measures to existing buildings. More specifically, the Metal Cladding & Roofing Manufacturers Association (MCRMA) Technical Papers 1–12 listed in the Bibliography provide helpful guidance on the use of metal cladding and roofing to avoid problems in this area.

11.10 Summary

Most modern non-residential and many residential buildings have cladding as the main form of wall finish. Problems such as water ingress as a result of joint failure or surface degradation are likely to continue to pose as the main defects in claddings.

Chapter 12

Doors and windows

OVERVIEW

This chapter deals with windows and doors, the two main external fabric components that have moving parts. It addresses the typical problems associated with these two important products.

12.1 Background

Decay of timber external joinery by wet-rot fungi has been widely reported in recent years and has affected many comparatively new properties (Figure 12.1). Both rain and internal condensation have been the sources of moisture which caused decay. Rain penetration at window and wall intersections has not been uncommon.

12.2 Doors

A wide survey of decay in doors, carried out in the United Kingdom, showed that a high percentage of unprotected external panelled doors, aged 12 years or less, were affected (Savory and Carey 1975). Entry of moisture has occurred mainly at joints and, with glazed doors, has soaked into the framing where putty or glazing beads have come away from the glass. Such beads frequently offer little protection. A contributory cause of rain penetration has been inadequacy of the tenons of the bottom rail, leading to slight dropping of the rail and consequent exposure of the joint between the rail and the panel (Figure 12.2). The absence of a weatherboard on the outer face at the bottom of the door has meant that rain is not thrown clear of the gap under the door. This problem has often been made more serious by the omission of the weather bar from the sill or its incorrect positioning.

Entry of water at the joints can lead to failure of the urea formaldehyde glue usually employed. It can also promote differential moisture movement where the grain of the timbers which meet at the joint runs in different directions. The stressing of the joint through moisture movement, and its

Figure 12.1 Wet rot in window frames.

weakening through failure of the glue, leads to loosening of the joint and an increased opportunity for moisture to enter and cause decay. Delamination of plywood panels in external doors occurs when internal-grade plywood has been used. The outer ply becomes wrinkled and often shows signs of splitting at the edges. This can be minimised if water and boil-proof plywood is used for exterior finishes. Plywood, no matter what its grade, is at risk from decay if moisture can enter through unpainted end-grain.

Distortion of doors is also likely if moisture penetrates and may occur, too, if humidity and temperature conditions are markedly different on the two sides, as they often are for an external door, and the door is poorly protected from such changes by the lack of an adequate paint treatment. Doors may then jam or give a poor fit in the door frame.

12.3 Windows

12.3.1 Background

Similar, but more extensive, problems have occurred with many timber windows. Modern machine-made joints are more complex than the older mortice-and-tenon joints, which were primed before assembly, and often

Figure 12.2 Rain penetration at window/wall joint.

result in large surface areas in contact. The adhesive used is not always applied in sufficient quantity, and with sufficient control, to ensure that all such surfaces are properly covered, and some are left as comparatively easy routes for moisture to penetrate. In any case, the adhesive does not seal the surfaces as effectively as primer. Some window frames used have been insufficiently robust and have distorted through moisture movement, this distortion then permitting the easier entry of moisture.

Moisture may find its way into window frames for a variety of reasons. Unfortunately, apart from the modest amount given by the manufacturer's primer, protection during transit, on site, and after installation, is frequently poor or even totally absent. The moisture picked up by timber windows as a result can be trapped when the frame is painted prematurely. Moisture may also enter through contact between frames and wet masonry. Current practice is to set the window frame in the outer leaf, which will be wet for long periods of time, and the frame itself, being so far forward, will also be more exposed to rain and sun. Rain may find its way directly behind glazing putties, which have moved away slightly from the glass, and into the joints of joinery when these have not been properly protected by paint. Rain will also collect on the horizontal surfaces formed if top-hung or centre-pivot lights are left open. An important route for entry, and one which seems to be increasing, is through condensation on the inner side of the glass. Condensation has been dealt with in some detail in Chapter 7. It may suffice here to state that condensation running down the pane collects on the usually rough and horizontal surface of the back putty and may remain there for long periods, slowly finding its way into the frame. Once it has entered, subsequent evaporation may be slow especially when, as is usually the case, both the internal and external faces of the frame are painted with a relatively impervious high-gloss paint.

12.3.2 DPC at window openings

Rain penetration at the junction of window frames with walls has been due to inadequacies in the detailing of the DPC at jambs, heads or sills. Windows are usually positioned as the wall is being built, and this is preferable to fixing the window in an already formed opening, for it is then much more difficult to ensure an effective barrier to rain. The window frame can be used to close the cavity but often the frame is set towards the external wall face and the internal leaf of the wall returned to close the cavity. Failure commonly takes place at the jamb through reliance on a simple butt joint between the vertical DPC and the frame, and also through failure to extend the DPC into the cavity (Figure 12.3). A simple butt joint is most unlikely to provide an effective barrier, even in sheltered conditions. Any shrinkage away from the wall will leave a gap for rain to pass directly between the frame and the wall if the junction there has not been well pointed with a mastic sealant. Even if it has, the DPC can be bypassed if the construction is such that internal plaster can be directly in contact with the external brickwork at the jamb. This can happen when the inside face of the frame does not extend beyond the inside face of the external

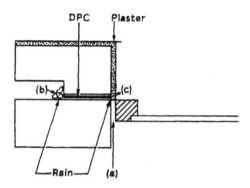

Figure 12.3 Rain penetration at window jamb. Poor practice: (a) gap, caused by shrinkage of frame from wall, not sealed with mastic. (b) DPC, not projecting into cavity, can be bridged by mortar droppings. (c) DPC bridged by plaster if inside face of frame is flush with inside face of external leaf.

brickwork. Rain penetration from this cause will be particularly likely if strong, relatively non-porous bricks, unable to absorb rainwater, are used for the external leaf. Wrong positioning of the vertical DPC with respect to the frame, allied to lack of, or inadequate, mastic pointing, is a common cause of rain penetration and one which should be suspected as the most likely when it occurs.

The details to be recommended will vary with the type of window, its relationship to the face of the wall and whether it is to be built in as construction proceeds or placed later in an opening. A positive seal can be achieved by having a recessed, or grooved, frame and projecting the DPC into this. The frame can also be kept out of direct contact with the brickwork. The DPC should then be nailed or stapled to the frame before the window is built in. The DPC will then need to be wider than the normal half-brick width, but a wider DPC is needed, anyway, to ensure that it extends at least 25 mm into the cavity to prevent its being bridged by mortar droppings (Figure 12.4). Where joints in vertical DPCs are necessary, the upper piece should lap to the external face of the lower piece. Care should be taken to see that any mastic used for sealing can take the strain which will be caused by differential movement between the window frame and the wall. The correct joint width for the sealant used needs to be achieved (see Chapter 11). If the joint to be sealed is open at the back, a back-up strip, usually of foamed plastics, needs to be used. This will help to ensure that the mastic sealant is forced against both sides of the joint and thus gives a good seal.

It is common practice to stop sills just short of the wall and, in theory, this could be useful, as it prevents the ends from being in direct contact

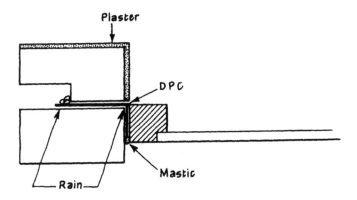

Figure 12.4 Better practice at jamb. DPC projects at least 25 mm into cavity and is tacked to the window frame. Mastic adds to the weatherproofing.

with wet masonry. However, if the joint is unfilled, rain can easily penetrate it and may possibly reach the interior. It will undoubtedly reach the end-grain of timber sills. The joint should be well filled with mastic. Rain can also penetrate at the bed joint beneath the sill when the latter does not have an adequate projection beyond the face of the wall, when the slope is shallow and when the drip is poorly formed or omitted. Many timber sills are constructed from two separate pieces, for this enables smaller and, thus, less costly pieces of timber to be used. It is essential that the outer piece sheds rain quickly and in no way enables water to be retained or fed back to the inner section. Examples are frequent where this has not been the case.

Masonry or rendering underneath a sill will be damp and may be excessively so when the sill projection is inadequate. Good practice requires a DPC below the sill to prevent such dampness from rising to reach the sill, and this will need to be detailed carefully to link with the vertical DPC at the jamb. The sill DPC will need to be turned up to form an upstand and the jamb DPC to overlap it so that rain is shed out and not in.

The lack of condensation channels on the inside surface of window sills has been lamented already. The need to slope the surface of the sill here is as necessary as it is on the outside, so that condensation does not feed into the window assembly, be it timber or metal. Too often, nowadays, the sill detail approaches that shown in Figure 12.5 rather than that desirable (Figure 12.6).

Setting windows back from the face of the wall helps in giving the window assembly some protection from the weather but will usually necessitate the use of a sub-sill to throw rain clear of the wall. The top surface of the sub-sill will receive run-off from the glazing and will itself be exposed to the weather. Water penetration between the sill and the sub-sill can occur

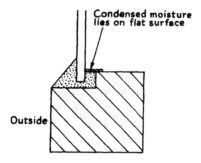

Figure 12.5 Damage by condensation on the inside of the window frame.

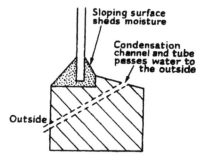

Figure 12.6 Removal of condensed water.

unless detailing is good and construction properly undertaken. The use of a weather groove in the sill is a common method of preventing such penetration but this needs to be positioned accurately so that any water penetrating drips into the cavity of a cavity wall and not straight onto the inner leaf (Figure 12.7).

The causes of rain penetration at window heads are similar to those dealt with in Chapter 10. Failures have occurred through inadequate lapping of the materials used in the construction of the cavity tray at the window head and by damage to it. Mortar droppings, too, have led to bridging of the DPC and to restriction of discharge of water through weepholes. Poor detailing at the junction between head and jamb has allowed rain to penetrate to the inner leaf at the ends of the lintel. Where the inner leaf is of block, failures have occurred because the normal depth of flashing of 150 mm is insufficient to reach, and to be tucked into, the first blockwork joint. This difficulty can be overcome by using either a flashing of greater depth or shallower blocks just above the lintel. The cavity tray should extend beyond the edge of the jamb DPC and the toe of the tray should project past the window head.

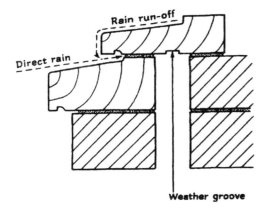

Figure 12.7 Sill must be accurately placed so that weather groove is over the cavity and not the inner leaf.

The pressed steel lintel, in most cases, acts also as the cavity tray and is available in lengths which obviate the need for jointing – always a source of weakness. The lintels are of galvanized steel and other protective coatings may also be applied: so far, chlorinated rubber paints and thermosetting epoxy powder coatings have been used. Provided that the long-term durability of the protected steel is satisfactory, this development seems to have promise in helping to prevent many of the problems associated with the separate introduction of a flexible cavity tray at window heads. The use of steel lintels may give rise to cold bridging and condensation (see Chapter 7), but some are designed to be plastered with an insulating plaster on the portion which is supported by the internal leaf and this should go some way to mimimizing the difficulty. Metal window frames have a high thermal conductivity and provide a cold bridge between the exterior and interior. This can lead to condensation problems, too, which can be quite severe. Insulation needs to be added to the frame to reduce the problem and this can be nearly impossible after installation. It may be noted that, while double glazing can reduce the risk of condensation on the glass itself, it will not necessarily prevent all condensation there: this will depend upon the internal humidity and temperature. Double glazing will not, of itself, affect the likelihood of condensation on the frame.

12.4 Prevention of failure and remedial work

The presence of decay in the woodwork of doors and windows is readily detected. The timber surface is often dished through the falling away of the soft and friable underlying timber. Paint surfaces are generally cracked and peeling away from the underlying wet timber. Putty is usually loose, having shrunk away from the timber, and pieces are cracked or missing. Replacement of putty is comparatively straight-forward but, whether or not it is feasible to cut out any decayed timber and renew it with preserved timber, clearly will depend upon the extent of the decay. Total replacement may be necessary. Where decay is localised, it may be possible to cut out the affected wood, to apply liberally a suitable preservative and to replace with treated wood. It will be necessary, generally, to remove and replace the putty and paint, and to fill open joints with a water-insoluble filler. The British Wood-working Federation has its own performance standard for wood windows, and included in the scope of this standard is the requirement for preservative treatment where necessary. In practice, this means that all softwood windows manufactured by its members are impregnated with preservatives.

Where DPCs and cavity trays are defective and have allowed rain to penetrate, the inner walls will show local patches of dampness after rain, adjacent to the defect, though these may be masked by decoration and the first signs of trouble may be after drying out, when efflorescent salts may appear. Remedial work is likely to be difficult and expensive, particularly to DPCs and cavity trays at jamb and head. Cleaning out mortar droppings from a flexible cavity tray after the mortar has hardened may be almost impossible without damage to the tray. Rain penetration may often be prevented or minimized by mastic pointing. Pointing is a simple operation and should be tried before replacement of existing DPCs is contemplated. It may also be possible, in areas where exposure is slight to moderate, to use a water repellent, despite the possible drawbacks mentioned in Chapter 10.

Specific detailed checks are strongly recommended at all parts of the DPC system before the work is finished. The cost of checking the initial design of DPCs at window openings and, carefully, the adequacy of site construction, is a small fraction of that which will be necessary to open up and replace the DPC later. This is true even at ground level in simple dwellings, let alone for more complex buildings at a height.

A thermal expansion problem can arise with windows made from uPVC. Dark colours can give rise to dimensional instability and enhance the risk of cracking of the glass. This is because the material naturally has a high coefficient of thermal expansion and dark colours will lead to greater temperature changes. It is safer to choose white frames and to keep them clean to avoid the build-up of dirt.

Finally, it should be noted that external doors and windows need to be selected to fit the probable exposure conditions and not bought on grounds of lowest possible price. Windows, in particular, are very exposed and are often not easy to replace. A little extra cost here could be money well spent, though even a better-quality window will need far better handling and storage on site than has been customary in recent years.

12.5 Summary

Windows and doors, unlike most other building components have moving parts. As a result they are subject to more wear and tear and are vulnerable to storm damage as well as damage by intruders. Inevitably the service life of these components is often limited.

Chapter 13

Roofs

OVERVIEW

This chapter considers the main problems associated with what is probably the most vulnerable part of a building – the roof, which is fully exposed to the vagaries of the weather. It analyses the common failures in both flat and pitched roof construction.

13.1 Background

 As a generalisation, it can be said that pitched roofs have given few problems and flat roofs have given many. The best advice one can give is to use a pitched roof wherever possible. However, post-war design has resulted in many buildings, particularly offices and schools, being of wide span, which has undeniable advantages. To top a wide-span building with a pitched roof would often mean that the total roof height would be unacceptable in terms of both visual appeal and first cost. For such a building, a flat roof may be the only feasible solution. To use flat roofs for buildings of comparatively small span, however, such as dwellings, is to court disaster, as many local authorities have found to their cost. The chances of success over a long period are minimal, given current standards of construction and the heating and ventilating regimes likely in dwellings.

13.2 Flat roofs

13.2.1 Preamble

Flat roofs fail because they let rain through; construction water is trapped which afterwards leaks out; or moisture generated within the building condenses and drips back. Moisture from the last two causes can also assist in the breakdown of the waterproof covering which, in turn, can lead to rain penetration. With some roof decks, the continued presence of moisture has been a contributory factor in structural failure. A major cause of leakage has been an insufficient slope to the roof. A designed fall of 1 in 80 does not result in a finished fall of that gradient. Variations in constructional accuracy, settlement and thermal and moisture movement lead to lower finished falls. Local areas of the roof, too, can deviate markedly from the overall falls, particularly around features such as roof drains. Local ponding, rather than shedding, of rainwater then occurs, which in turn can lead to further local deflections, more ponding and slower drying (Figure 13.1). Rain leakage through roofs following splitting and deterioration of the waterproof coverings has been due, in part,

- to failure to recognise, or to allow for, the differential moisture and thermal movement to which a flat roof system is prone;
- to the adverse effects of standing water and solar radiation; to blistering caused by pressure of entrapped moisture or air;
- to inadequate detailing at parapets and projections; to decay and collapse of some materials used as decking or for insulation; and,
- to mechanical damage.

If a decision has been taken to use a flat roof, then the designed falls should be at least 1 in 40 and preferably achieved by sloping the structure rather than by forming them by a screed or the insulation. There are many types of flat roof and waterproof covering (Douglas 2006). There are, however, some issues which are common to much flat roof design and construction, and these are now considered.

Figure 13.1 Ponding on flat roof – a state typical of many flat roofs.

13.2.2   Dripping of moisture

Water used in the construction of concrete decks, and in lightweight concrete screeds used to provide falls and as thermal insulation, can be slow to evaporate. In the United Kingdom, an unprotected deck or screed is unlikely to dry out, except in a long spell of dry, warm weather. Only in the driest parts of the United Kingdom, in fine weather, does evaporation exceed average rainfall and it is seldom possible to apply waterproof coverings to dry concrete roof systems. A common failure has been the dripping of water, usually stained brown, from the underside of the structural roof and, particularly, from natural outlets, such as electrical conduits. Very often, the cause of this dripping is taken to be rain passing through a defective covering. This may be the case but it is more likely to be caused by entrapped water used during the construction of the deck or by entrapped rain which fell during construction. As mentioned, most roof-laying operations in the United Kingdom will be affected by rain. In some cases, this may not matter, because the water will drain through the roof and evaporate more or less harmlessly from the underside. However, when a vapour barrier is used beneath the roof deck, rain may be trapped between the vapour barrier and the waterproof covering. It can remain trapped for some years, very slowly migrating towards low points in the structure and emerging wherever it reaches some feature which has resulted in a discontinuity in the vapour barrier. It then drips out. When a vapour barrier has not been used, moisture derived from internal activities may condense beneath a waterproof covering, particularly, though not exclusively, in the winter and can drip back in a similar way.

It may be difficult to decide the exact cause of dripping, but that originating from trapped rain or construction water will have no marked seasonal tendency to manifest itself, while that due to condensation is more associated with the winter. If it has been through leakage of the waterproof covering, then it will be associated with rainy periods though not necessarily coincident with the rain. The position at which drips appear can, of course, give some indication of the cause. Thus, defects in parapets, skirtings and verges are likely to result in drips near the junction of the ceiling and external walls. Defects caused by movement at construction joints may well give rise to drips near the junction of ceilings with internal loadbearing walls.

Any entrapped moisture needs to be drained away. Drainage can be affected by puncturing the underside of the roof at the low points. The holes are closed later when the base has dried. Ventilators passing through the waterproof covering may also be of value in helping to release moisture-vapour pressure, though ventilators will not, of themselves, dry out the screed or base. No vapour barrier should be used which could seal in such water.

13.2.3 Warm-deck and cold-deck roofs

Condensation in flat roofs, whether they are covered with asphalt or are constructed in other ways, can be prevented or reduced by providing an effective vapour barrier at the warm side of the roof structure and by ventilating the roof to the outside air. In the case of a flat concrete roof, condensation can be prevented, in general, by the provision of insulation above the structural deck and separated from it by a vapour barrier. The latter should be of high quality: the view had been expressed (Department of the Environment 1981) that bitumen felt types 1B or 2B of BS 747 are scarcely adequate and some holding Agrément certificates have been preferred. The roof described is known as a warm-deck roof because, in a heated building, the roof deck will be warmer than it would be if the insulation were provided underneath the deck. With insulation above the deck, the latter, and the ceiling below, will be at a temperature close to that of the interior of the building. The insulation is generally protected at its upper surface by the waterproof covering. In a warm-deck roof, it is not necessary to provide cavities in the roof system specially for ventilation except in cases such as heated swimming-pools where pressurisation with fresh air may be necessary. However, a warm-deck roof will have a cavity formed between it and the ceiling when this is of plasterboard on battens. Such a cavity is of value if the building is intermittently, rather than continuously, heated.

If the insulation to be used is, itself, not affected adversely by water, either in relation to durability or the retention of its insulating properties, then it may be placed not only above the deck but also above the waterproof layer. The commonest of such insulating materials are foamed glass and extruded polystyrene. This type of roof construction is, commonly and rather confusingly, termed 'inverted' warm-deck roof construction. A better name, also used, is a protected membrane construction. Both the protected membrane construction and the conventional warm-deck roof construction will pose different constructional problems. In the case of the protected membrane construction, the waterproof layer is protected from sunlight, from large thermal movement and from any traffic on the roof. Some limited experiments over a 6-month period have shown that temperature fluctuations in asphalt can be as little as one quarter that of asphalt placed over the insulation and the rate of temperature change very much slower (Department of the Environment 1979). A protected membrane roof also avoids the need for a separate vapour barrier, as the waterproof layer performs this function itself. Construction moisture is not trapped and can dry out downwards. However, the insulation will need to be held down against wind forces, drainage design will be more demanding and the water-proof layer may not be accessible to any repair needed without partial destruction of the insulation. It will be necessary to weigh down the insulation by gravel, by slabs or some other means as the work proceeds to prevent it from being blown away. In conventional warm-deck roofs, where the insulation is below the waterproof covering, the latter is readily accessible for maintenance but is subjected to much thermal movement. If leakage does occur, water may be trapped in the insulation and this can reduce its insulating properties. It can cause rotting if the insulation is cellulosic. A separate vapour barrier is required below the insulation. Good solar protection to the waterproof covering will also be needed. The covering will also need to be protected against thermal movement at the joints between rigid plastics insulating boards.

In cold-deck roofs, thermal insulation is provided below the roof deck (see BRE Digest 312, 1986). A ventilated air space will be needed between the ceiling and the deck to reduce, or prevent, condensation, and this can pose many difficulties in design and construction. While it may be possible to design and construct cold-deck roofs to prevent general condensation, it is difficult to prevent localised condensation and the risk of failure is greater than that for warm-deck roofs. It is for this reason that cold-deck flat roofs are no longer recommended anywhere in the United Kingdom. In fact their use has been banned in Scotland for many years.

Protected membrane construction offers the balance of advantage for heavy decks of concrete. It may not, however, if it is of lightweight construction and particularly when it is of troughed metal. It has been shown (BRE Annual Report 1981/82) that heavy precipitation during cold weather can reduce the effectiveness of the insulation and can lead to a significant fall in temperature at the underside of the deck. This may be sufficient to cause condensation there. This risk should be considered consciously before the decision is taken to place the insulation above a lightweight metal deck. It is wiser not to do so in wet, cold areas of the country.

13.2.4 Asphalt coverings

Asphalt is an inherently durable material but can crack if subjected to a sudden stress, particularly at low temperatures. The most common causes of cracking are movement of the deck upon which the asphalt has been laid and differential movement between the deck and the features such as skirtings, parapets, verge trims and flashings.

Thermal and moisture movements of materials have been considered already and individual values need not be reiterated. It is relevant to note, however, that a flat roof is very exposed and subjected to wide variations in temperature. Asphalt has a high coefficient of thermal expansion and, being black, is likely to be subjected to large temperature changes unless specially protected from solar radiation.

It is probably well known that an isolating membrane needs to be interposed between the asphalt and the deck it protects, but failures have happened from time to time because of its omission. Mostly, the omission has been partial, a typical case being that where the membrane has been stopped short of the verges of the roof deck and the asphalt taken beyond the membrane and bonded directly to the deck at the verges. It is less well known that cracking of the asphalt may also be induced if a membrane is not also interposed between the asphalt and any paving tiles or sand/cement screed placed on top as a surface finish, or if too thin a membrane, such as polyethylene sheet, has been used. The detailing of asphalt over expansion joints in the roof deck has often been deficient, when a flush finish is required rather than the twin-kerb recommended in BS 8218. Failures have been due mainly to separation of the asphalt from the flanges of flush-type proprietary expansion joints. Partly, this has been through disregard of manufacturers' instructions but the placing of asphalt in conjunction with the materials and joint profiles used to form the joint can be difficult.

Poor detailing at parapets and projections through roofs has been a cause of many failures. Extensive cracking has occurred where asphalt has been dressed up parapets to form a skirting and tucked into chases, without allowance in design for the differential movement between deck and parapet. This has been associated, particularly, with timber and wood-wool decks. Many cases of rain penetration behind the asphalt skirting at the parapet junction have been reported, even when cracking has been avoided. Penetration has been due mostly to slumping of the asphalt from the chase into which it was tucked. The main reasons have been that the asphalt taken up on the parapet and into the chase has been too thin; the solid angle fillet of mastic asphalt formed at the junction of roof and parapet has been too narrow at the face; the vertical surface has been too smooth and/or wet to allow the asphalt to adhere properly; and the chase, itself, has been of insufficient size. The lack of a good solar-protective finish will contribute to sagging and blistering. When the horizontal chase is too small, it gives little support to the asphalt where the latter is tucked in and leaves insufficient room for the cement/sand pointing which is required above the asphalt. If the asphalt slumps and shrinks away, rain will penetrate eventually behind the skirting and into the roof deck. The finishing of flat roofs at the edges using aluminium trim into which the asphalt is dressed has, on the whole, not been very successful and cracking has occurred, particularly at changes in direction of the trim. This, too, is basically because of differential thermal movement, leading to fatigue.

Severe cracking of asphalt has been induced when ordinary emulsion paints have been used as reflective treatments to reduce the absorption of solar heat: such paints can cause deep cracking. Asphalt may also crack if overheated during laying, though this is not usually of major significance unless heating has been grossly excessive. Surface crazing is fairly common, and many cases occurred in the fairly hot summer of 1975, with crack widths up to 6 mm. Crazing is due, in general, to failure to sand-rub the asphalt with a wood float.

Sand-rubbing not only smoothes out imperfections in the final surface but it helps to absorb bitumen brought to the surface during laying of the asphalt. Crazing can be due, also, to the lack of a good surface reflective treatment and to ponding of water on the roof which can lead to large temperature gradients and, thus, to strain between dry and wet areas. Surface crazing, however, is seldom of great importance and does not lead directly to rain penetration though it may reduce the total effective life of the asphalt. Asphalt may blister through pressure generated by water vapour, particularly in hot weather, when the pressure can be high. Moreover, in hot weather, the asphalt will be softer and less able to resist such pressures than it would in cold weather. When the roof is of concrete and, particularly, when a lightweight concrete screed has been used, the source of the moisture is mostly the water used for construction. This can be trapped when the asphalt is laid and escape will then be slow: rain may also be trapped. Water may collect, too, as a result of interstitial condensation and may be sufficient in quantity to cause blistering, though this is more likely with built-up felt roofing than with asphalt. Blisters in asphalt can be a sign of omission of, or gaps in, the isolating membrane. Blisters may be of many sizes: diameters of 150 mm are not uncommon. Because the surface area of the blister is greater than that of the corresponding asphalt before blistering, it follows that the asphalt will be thinner than normal over the blister and it may become sufficiently thin to split. It will then, of course, not prevent rain from entering. Many blisters, however, do not split, and do not allow rain penetration, though they will be prone to damage by any foot traffic on the roof and may enhance local ponding and, thus, general deterioration (Figure 13.2).

It is, of course, necessary that the deck to which it is applied remains sound and provides a stable support to the asphalt. This has not always been the case, particularly when asphalt has been applied to wood-wool slabs. Failures have been caused by the seriously different levels of adjacent slabs and by excessive gaps between slabs. Differences in levels of as much as 18 mm, and gaps of 13 mm, can occur. These uneven surfaces have resulted in the asphalt also being uneven and forming into ridges between slabs leading, in turn, to ponding and cracking. Once wood-wool slabs under asphalt become wet through leakage, the moisture does not escape readily and persistent dampness causes decay of the slab, general loss in strength and disintegration. This leads, in turn, to further cracking and disintegration of the asphalt. A further disadvantage of wood-wool is its high moisture movement. Failures have also been caused by the distortion of some forms of chipboard. Chipboards based on urea-formaldehyde as the binder are highly susceptible to the ingress of water for whatever reason, and when they become wet for any reasonable length of time, they weaken and distort. This leads to distortion and failure of the asphalt. Inadequate support of wood-wool slabs, chipboard and strawboard leads to sagging,

Figure 13.2 Blistering of asphalt.

loss of support for the asphalt and its eventual fracture. Boarded timber decks tend to warp and joints to open, and these have provided an unsuitable substrate for asphalt though, given a high standard of site supervision and construction, satisfactory roofs have been achieved. Asphalt does not take kindly to damage typically caused by careless handling of scaffolding, the use of concrete mixing plant, bricks and tiles thrown down upon it, spilt paint and solvents, and the temporary storage of heavy equipment which can cause local deformation and indentation. Damage has also been caused by the movement of metal fixings, such as cradle bolts, poorly fixed to the structural deck beneath.

There are a number of actions needed to avoid these failures of splitting, or disintegration, of asphalt, and of leakage. It is first necessary to ensure that differential movement between asphalt and the roof deck and contiguous features, for example parapets, is minimised. A black sheathing felt, type 4A (1) of BS 747, should be used as the isolating membrane between asphalt and deck. Where the deck and any parapet are likely to experience markedly different thermal or other movement, it is better to apply the asphalt to a freestanding kerb, fixed to the deck, with a clearance between the kerb and the parapet to permit movement. The gap between the kerb and the parapet should be covered with a metal flashing (Figure 13.3).

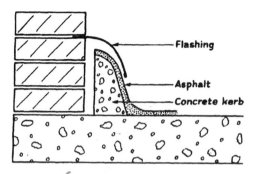

Figure 13.3 Separation between roof deck and parapet. Free-standing kerb needed when deck and parapet are likely to be subject to markedly different movements.

Where the parapet and deck are not subject to widely differing movement, the asphalt should be tucked into a chase in the parapet. The chase should be not less than 25 mm × 25 mm in cross-section, the exposed part of the asphalt should be splayed to shed rainwater and cement/sand mortar used to point between the top of the asphalt and the underside of the chase. The asphalt will need to be applied in two coats to a total thickness of not less than 13 mm and the solid angle fillet of asphalt formed at the junction of parapet and deck should be not less than 50 mm wide on its face. The height to which the asphalt is taken up the parapet should be at least 150 mm. Smooth and wet surfaces must be avoided and keying to concrete parapets may be necessary, particularly if the concrete was placed against metal or plywood shuttering. Expansion joints are required in the asphalt to coincide with expansion joints in the structure and should be of the twin-kerb variety, if at all possible, for flush expansion joints are very difficult to make waterproof.

If wood-wool slabs are to be used as the deck, it is essential that they are firmly fixed to their supports and are covered by a screed at least 25 mm thick. Adjacent slabs should vary by less than 3 mm in level and should be jointed firmly and taped with scrim. Chipboard should not be used for flat roof decking where occupancy conditions are likely to give rise to high humidity, for example in laundries. It should be selected and applied as recommended in BS 5669.

The safest solar reflective treatments are mineral aggregates, light in colour and set in a bituminous compound; tiles of asbestos cement similarly bedded; or a sand-and-cement screed cut into paving squares, provided that this is separated from the asphalt by a layer of building paper or similar membrane. Some bituminous aluminium paints may be satisfactory but manufacturers' advice should be sought and followed closely.

Major cracking of asphalt through roof movement implies that separation between the two has not been effected by the proper use of an isolating membrane, and complete replacement may prove necessary. It may be possible, however, if cracking is along one or two lines, to provide movement joints of the upstand variety at these points. Splitting in the region of the angle fillet at parapets will need to be remedied by removal of the asphalt over the full vertical height of the skirting, which should be at least 150 mm, and removal horizontally for about the same distance. A free-standing kerb will need to be provided, with a 13 mm gap between it and the face of the parapet, and detailed as recommended in BS 8218 (see Figure 13.3). Where rain penetration has occurred behind the asphalt at skirtings through inadequacies in the chase, it will be necessary to remove the pointing and the asphalt for a distance of some 60 mm below the chase and to reform the latter, apply fresh asphalt and repoint. Asphalt which has cracked through the application of a harmful reflective paint is likely to need complete replacement. Blisters on asphalt, which have not split, may be left but inspected periodically: they are not necessarily important. However, those which have split will need to be completely opened but can then be patched locally. The removal of entrapped water can be accelerated by dewatering, using a suction pump. This is likely to involve removal and subsequent replacement of the asphalt.

13.2.5 Built-up bituminous felt roofing

Common defects in built-up felt roofing are splitting, blistering, ridging and rippling, local embrittlement and pimpling, and loss of grit. Splitting is caused mainly by excessive differential movement between the felt and the substrate to which it is attached. Felts may be based on organic, asbestos or glass fibres but none can be stretched, without splitting, by more than around 5 per cent, and less if the felt is aged. Differential movement which could cause such an extension occurs commonly. It is because of this that the lowest layer of the usual three-layer felt system is recommended to be only partially bonded to substrates likely to impose undue stress on the felt system. These include concrete and screeded surfaces, screeded wood–wool slabs, particle boards and laminated boards. Extensive movement of the substrate has been due either to its drying shrinkage or through poor fixing to primary supports, such as timber or steel joists. This has been particularly the case with chipboard and with wood-wool. At one time it was believed that felt applied to expanded polystyrene boards should be fully bonded to them for, as with cork and fibreboard, no major differential movement was expected and, hence, no need for partial bonding of the lowest layer of felt.

However, many cases of splitting of felt occurred and this is now thought due not to movement of the polystyrene, but to excessive thermal movement of the felt, caused by the high degree of insulation provided by the polystyrene. This will lead, both annually and daily, to wide variations in the temperature of the felt. The repeated and excessive thermal cycling of the felt has led to its fatigue failure. Some failures, when expanded polystyrene boards have been the substrate, have also been caused by partial melting of the boards through the direct application of hot bitumen on them, used for bonding the felt. This has led to depressions and to poor support for the felt where it bridges over them. Splitting commonly takes place at joints and this can happen even when the felt is only partially bonded, if the width left unbonded is too narrow to take the strain imposed by the joint movement. As with mastic asphalt, built-up felt roofing will split if the substrate fails to provide an adequate base, through excessive distortion caused by lack of strength. Both wood–wool slabs and urea-formaldehyde chipboard are likely to lose strength if they become, and stay, wet, as might happen by leakage of the felt covering, by condensation or by entrapped moisture.

Blisters occur far more readily in built-up felt roofing than they do in mastic asphalt (Figure 13.4). They form either between layers of felt, commonly under the top layer, or between the felt and the deck or the insulation, if the latter is used above the deck. In both cases, they are due to the expansion of entrapped air or moisture by solar heating. However, blisters should not be taken necessarily as a sign of failure. They are common in

Figure 13.4 Blistering and ridging of bitumen felt.

practice and often have little significance though, like blisters on asphalt, they present a thinner surface and one, therefore, more prone to damage. Long undulations or ridges may form in felt to give a rippled appearance. In part, this may be due to storage of felts flat, instead of upright, which can cause permanent distortion. In use, there are two main causes. If the ridges are roughly parallel and extend for a considerable distance, the cause is probably their coincidence with gaps and unevenness in level between boards or slabs such as insulation boards or wood-wool. These ridges are essentially hard when pressed. Others, more yielding to pressure, are caused by failure to allow the felt to flatten properly before it is fixed, by poor distribution of bitumen compound used for laying, by insufficient pressure during laying and by expansion of the felt. The entrapped air or moisture expands by solar heating, particularly in summer months when the felt is more flexible and heating is greatest. As with ordinary blistering, ridges may have little significance but do tend to increase ponding and need to be avoided. Minor local cracking, pimpling and pitting, with embrittlement of the felt, is commonly seen. These minor defects are caused by the attack of ultraviolet radiation together with atmospheric oxidation of the bitumen. They are, to an extent, a normal consequence of ageing. This ageing is assisted by the loss of reflective grit which, in turn, will be caused by the slow decay of the thin layer of bitumen used to bond the grit to the top layer of felt. In warm, sunny weather, the more volatile fractions in the bitumen expand and lead to a slow crazing and, ultimately, to exposure of the fibres of the felt. Breakdown cannot be wholly avoided but is accelerated by standing puddles, by the wrong choice of reflective treatment or lack of its maintenance, and by the choice of the wrong felt for the top layer. In the early stages, crazing is superficial but, if not attended to, can deepen and lead, eventually, to failure of the felt to achieve its sole purpose of keeping out the rain. The loss of protective grit through wind and rain is inevitable. It is accelerated by poor practice, such as the use of too small a grit size, by application of grit to the felt in wet weather, by the use of bitumen emulsion for binding and by standing water where the roof has an insufficient fall. Loss of grit, apart from exposing the felt to aggressive solar radiation, may also lead to choking of gutters and rainwater outlets which, in turn, results in longer periods during which water lies on the roof. This is particularly so when integral box gutters are used, for these often have only shallow falls because of the limited depth of the roof structure.

Poor detailing at upstands and angles, particularly at parapets, verges and rooflights, is a common cause of leakage of felt-covered roofs. Blistering, sagging and splitting of felt at skirtings and parapets occur for the same reasons as they do with asphalt (Figure 13.5).

Cracks between skirtings and the main roof covering are common and are due mainly to differential movement, enhanced by ageing of the felt and the

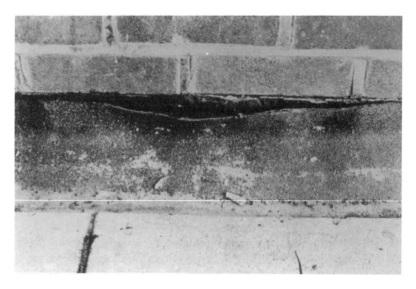

Figure 13.5 Splitting of felt upstand.

lack of a maintained protective finish. Sometimes, failure has been caused by the simple dressing of the felt into a chase and the omission of a separate flashing. At roof edges, good detailing is necessary to resist uplifting by wind. Metal edge trims have been used to give a neat appearance but are incompatible in thermal movement with the felt. Failure has been caused, either by breakdown in the bond between the felt and the trim or by splitting of the felt at the joints between adjacent lengths of trim.

As already mentioned, the effects of differential movement between the substrate and the felt are nullified by partial bonding of the lowest layer of felt and this practice needs to be adopted on concrete and screeded roofs, particle board, laminated boards and asbestos decks. It is essential that timber, chipboard and wood-wool slabs are firmly fixed as detailed in BS 6229. Wood-wool is used normally as a lightweight structural deck and to provide some insulation. It is not recommended for use where humidity is likely to be high. If used, the slabs should be pre-felted and left unsealed on the underside. Special methods are necessary when built-up felt roofing is used over expanded polystyrene. The latter should be at least high-duty grade in accordance with the requirements of BS 3837, and should be pre-felted with an underlay complying with BS 747. The underlay felt should be fully bonded to the polystyrene. A vapour barrier of bituminous roofing will be needed between the roof deck and the polystyrene, and a coat of hot bitumen should be applied to the top surface of this vapour barrier, allowed to cool to around 80°C, and the polystyrene boards then placed in position with staggered end joints. Fibre insulating boards, 13 mm thick, should then be coated with hot bitumen and pressed, bitumen-coated side down, onto the polystyrene, with joints staggered to break bond with all joints in the polystyrene. The built-up felt roofing is then fully bonded to the fibreboard by the normal technique. A warm roof design is preferred and, if dry insulating materials are used, a vapour barrier should be placed between the roof deck and the insulation. No such barrier should be used beneath a wet screed. If a cold roof is used, a ventilated air space will be needed between the insulation and the deck, and a vapour check should be fixed beneath the insulation. All uncompleted work should be covered from the weather. Felts should have side laps of 50 mm and end laps of 75 mm, with the lap joints arranged so that they do not impede roof drainage. Minor movement joints should be provided at lines where movement of the substrate is expected. Felt used for the top layer will need protection from solar radiation. The choice will usually lie between self-finished mineral-surfaced felts of type 2E or 3E of BS 747 or by mineral chippings, 6–13 mm in size, secured on site to the felt, using a solution of bitumen in a volatile solvent – not a bitumen emulsion or a hot-applied bitumen. The use of chippings makes any leaks hard to find and does lead to choking of gutters. Some reflective paints may be better but reliable data on performance and durability are lacking, and would-be users should be careful to seek manufacturers' advice. Ordinary emulsion paints should not be used. Where continuous foot traffic is expected, a more durable finish is necessary, as recommended in BS 6229. At upstands, the felt should be taken over an angle fillet and bonded with hot bitumen, to a height of at least 150 mm above the roof (Figure 13.5). A felt or metal flashing tucked into a chase not less than 25 mm deep will be needed to complete the weathering protection. At roof edges, experience has suggested that welted bitumen-felt trim gives a better protection than metal trim. If aluminium trim is used, it should be in lengths not exceeding 1300 mm, fastened at 300 mm centres, with a 3 mm expansion gap between lengths and covered on its top surface with a high performance felt capping. Where possible, gutters should be clear of the roof rather than integral with it. Bitumen felts are very sensitive to damage and no general building work should be allowed on the finished roof.

The common tendency is to become unduly worried about blisters and ridges, and it is often best to leave well alone but to keep a careful watch for signs of splitting. If repair is decided upon, it is best to do this in warm, dry weather. Blisters and ridges can then be cut out and new patches of felt applied. If blisters and ridges are numerous or extensive, it can be simpler to strip off the felt and relay. For minor cracking, it may be sufficient to apply a reflective treatment or a bituminous sealant. Deep cracking or major splitting will require felt to be stripped and fresh felt laid, with an appropriate reflective treatment.

13.2.6 Polymer roofing

In recent years, a range of polymeric materials, including notably poly-isobutylene, butyl rubber, PVC, chlorosulphonated polyethylene and polymer-modified bituminous roof membranes reinforced with polyester fleece, has become available. Many are designed for use as single-layer systems, stuck to the substrate with special adhesives and the joints between sheets solvent or heat-welded. Those used in single-layer systems require a high standard of workmanship and most are installed by approved installers. These products are more flexible than the conventional bitumen felts and retain good flexibility after ageing. The main problem has been water ingress through inadequately bonded joints and mechanical damage, causing splits. Such splits are caused relatively easily by stones, nails and tools dropped onto the polymers. Many of these systems have BBA certificates and have since the early 1970s demonstrated that they have considerable potential (see Douglas 2006).

13.2.7 Parapets

Parapets are severely exposed to rain and, also, to wide changes in temperature. They are, therefore, likely to suffer attack by sulphates and by frost. Brick parapets may also be subject to moisture expansion. The causes and symptoms of such attack and movements are considered in Chapter 10. Renderings on parapets are similarly exposed to severe weather conditions. Wherever possible, parapets should be avoided, but if a decision is made to have them, then cavity parapets should be adopted. With solid parapet walls, the problems of preventing rain penetration and the deterioration of any renderings used are such that failure is almost certain (Figure 13.6). All bricks used should be low in sulphates. In cavity parapets, failure has been promoted frequently by the lack of an effective coping and by ineffective DPCs in the parapet. The coping system needs to prevent the downward penetration of rain, it needs to throw any rain clear of the face of the parapet wall and the coping material itself must be of high durability to withstand the severe exposure. Brick copings are vulnerable to water penetration at the mortar joints and do not provide an angled weathering surface. In many cases, the bricks used have not had adequate resistance to either frost or sulphate attack and have disintegrated. In other cases, the lack of a DPC immediately under the coping, which is necessary unless the coping itself acts as a DPC (e.g. when it is of metal or plastic), has led to an unduly wet wall which is more susceptible to attack. A common fault with coping DPCs has been to stop the DPC short of the external face. Water which has passed through the coping, usually through the joints, can then pass into the brickwork below. There have been difficulties, too, when the DPC has been taken under the inner and outer leaves only of a cavity parapet wall and not right across the cavity as well (Figure 13.7). Rain can then still penetrate

Figure 13.6 Rain penetration through parapet wall.

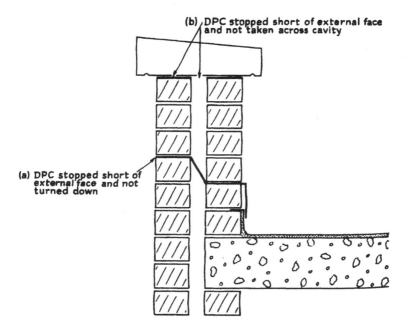

Figure 13.7 Common faults in parapet details: (a) Brickwork becomes wetter and more susceptible to frost and sulphate attack. (b) Rain can run down underside of cavity tray and into roof.

through the joints in the coping and drip down into the cavity, ultimately falling onto the brickwork below the DPC level. Unsupported DPCs taken right across the cavity are prone to damage and can sag, presenting some likelihood of separation at lap joints.

With cavity parapets, the coping should not be of brick. Units conforming to BS 5642 would be satisfactory. The coping should be weathered and throated, and its face should project at least 40 mm away from the face of the wall or rendering. The outer edge of the throating should be at least 25 mm away. An effective DPC should span right across the wall, immediately below the coping, should be well bedded between wet mortar and should be supported over the cavity portion. If parapet walls are kept low, it may be possible to take the waterproof covering to the roof up the wall and under the DPC, to form a continuous covering. Parapet walls of a height where this is not possible, because of limitations on the height to which the roof covering can be dressed safely, need a further DPC or cavity tray at the base of the parapet, above the point at which the roof covering is tucked into the parapet. If the DPC is missing or ineffective, then rain clearly can pass behind the edge of the waterproof roof covering, which will penetrate only some 25 mm into the wall. Many cases of moisture penetration are known to result from poor detailing of the junction between the DPC at the base of the parapet and the roof covering. A common error has been to stop the DPC short of the flashing which is commonly used to complete the waterproof detail at the roof junction. The cavity tray is usually arranged to slope towards the roof rather than towards the external face of the parapet wall and this has generally been accepted as the most effective form of construction, for it reduces unsightly staining of the external wall. However, it is most important that, in such a case, the tray is taken right through the external wall and turned down to form a projecting drip. If not, and because parapets are so exposed, there is a risk that rain will be conveyed down on the underside of the tray into the inner leaf and there bypass any flashing and the roof covering, and enter the structural roof. Damage can be caused to the DPC when this is located in the same bed joint as the chase cut to take a skirting. Unless considerable care can be assured, it may be better to place the DPC in the bed joint above and to provide a non-ferrous flashing running from the DPC, down the one course of brickwork and lapping over the bituminous covering. Weepholes are needed to assist drainage and, when the DPC slopes towards the roof, should be formed by leaving open perpends every fourth brick in the course of the inner wall of the parapet, immediately above the DPC.

Cracking and general damage to brick parapet walls by frost and sulphate attack will be recognised easily and will usually necessitate the rebuilding of the wall. Before so doing, it will be useful, however, to see whether it is possible to dispense with the parapet altogether. If not, it will be necessary to use bricks of low sulphate content and high frost resistance, and to keep the parapet as low as possible. Excessive moisture movement will result in over-sailing of the DPC at roof level. It may be feasible to rebuild only part of the wall and it will be an advantage to re-use the bricks, which will not display any further appreciable moisture movement. Thermal and moisture movement, frost and sulphate attack may also lead to open joints in coping units and to their distortion. In such cases, the joints will need to be raked out and repointed, and the coping replaced with appropriate expansion joints at maximum intervals of 9 m. Failure of the DPC/roof covering will show as dampness on the upper parts of the external walls and careful examination of the whole parapet system will be necessary to determine the cause. It is usually the DPC system at roof level which is at fault, through splitting or lack of overlap between the DPC and any flashing used. It may be possible to repair the system or to insert new flashings but complete rebuilding may prove necessary. In all operations, care needs to be taken to avoid damage to the roof covering and to provide temporary protection so that, during repairs, rain is prevented from getting under the waterproof covering.

13.3 Pitched roofs

13.3.1 Roof structures

Few problems arise with pitched roofs but, in recent years, some troubles have been experienced with overall instability of trussed rafter roofs. These are now used in probably more than 80 per cent of new domestic dwellings. Trussed rafters are lightweight truss units, generally spaced at intervals of 600 mm and made from timber members of uniform thickness fastened together in one plane, mostly by metal plates, though plywood connectors may also be used. Since their introduction into the United Kingdom in the mid-1960s, most have performed successfully, but there have been cases where trusses have collapsed or moved laterally whilst remaining parallel to one another – the domino effect (Baldwin and Ransom 1978). This has been caused by the omission of diagonal bracing and the fault is manifested initially by displacement of the roofing tiles. If not corrected, the failure can lead to rain penetration, buckling of rafters and cracking of gable walls. In some cases, failure of the whole structure has occurred through inadequacy of the roof/wall design. The bracing provided for stability of the rafter system, as such, was not adequate, nor designed, to provide stability against wind loading on gable walls (Figure 13.8). Often, the need for straps for lateral restraint between wall and roof has not been appreciated. The down-turned end of the strap is intended to bear tightly against the external face of the internal leaf, and the body of the strap itself to be fixed firmly to the adjacent truss ties. Common faults are that the straps are fixed to the ties

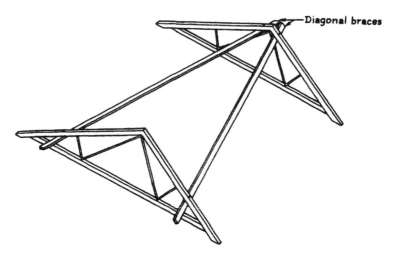

Figure 13.8 Diagonal braces under rafters, running from heel to apex, are needed for stability.

before the internal leaf of blocks reaches that level, and the down-turned end is kept clear of the future face of the blockwork to avoid fouling it. It then serves no useful purpose. In other cases, fixing to the tie is poorly done or omitted altogether (Bonshor 1980). Corrosion of galvanised steel fasteners has also been reported, which suggests insufficient long-term durability, though no failures from this cause are known as yet. The requirements for truss bracing are now included in BS 5268 Part 3. Installation of proper bracing at the time of construction is simple and cheap. The removal of tiles, battens and sarking, and the straightening or replacement of trusses which have moved laterally through failure to provide bracing will cost at least 50 times as much.

Trussed rafters are intended to be used under dry conditions, and need to be protected and handled with care during storage at the fabricators, during transit and on site. The rafters, when placed, should be covered with the minimum of delay. It is common, and bad, practice for these needs to be ignored, which can lead to distortion, damage to joints and corrosion of the metal fasteners (Figure 13.9).

13.3.2   Roof voids

Condensation can occur in pitched roofs as well as in flat roofs, and complaints have been made of condensed water running down rafters and even dripping through ceilings. There are several reasons for expecting that the incidence of condensation may rise in modern dwellings unless specific

Figure 13.9 How not to store trussed rafters.

steps are taken to avoid it. First, the increase in thermal insulation, with a minimum total thickness of insulation of 300 mm now recommended, will make roof spaces above the insulation colder and the dew-point more easily reached. Second, the placing of insulation in such thicknesses is likely to impede ventilation at the eaves and, if packed in right up to the eaves, will close the gap between the ceiling and the underside of the roof tiles. This will be so, particularly, when roof pitches are low which, again, is still a current trend. Chimneys, too, in many dwellings, are now either omitted or not used, and in such cases, there will be no heat gain from the chimney into the roof space to raise temperatures and so help reduce the risk of condensation. It is now also common practice to use impermeable sarking under the tiles. The main source of entry of water vapour into the roof space is from below, and ways of minimising the generation of moisture in buildings are described in Chapter 7. There are many routes by which moist air, however, can gain access to the loft. The principal ones include the access hatch to the loft, holes around pipes and ceiling light roses, and at the cavity wall heads, which are often only partially closed. Tests have shown (BRE Digest 270) that, typically, 50 per cent of the air passing through the ceiling into the roof space does so around the access hatch, 40 per cent through holes around pipes and most of the remainder through ceiling roses. In the loft itself, cold-water tanks may not have well-fitting covers and water can evaporate readily into the loft. This will be especially true when the water in the tank is warmer than the surroundings, as it will be when insulation is not placed under the tank – which, in itself, is good practice in that its omission helps to prevent the water from freezing. Condensation needs to be prevented as much as possible, and a primary reason in roof spaces is to keep the moisture content of the roof timbers at a level low enough to prevent decay (i.e. < 18%). It is also very necessary to prevent the metal plate fasteners in trussed rafter roofs from becoming wet. This is particularly true of the fasteners at the heel joint by the wall plate, which are vital to the structural integrity of the rafter system. This position is the most likely to suffer from lack of ventilation if insulation is packed tightly up to the eaves. It is very necessary to prevent such tight packing and insulation should never be taken right up to the eaves. On the other hand, there is the need to ensure that any insulation covers the whole ceiling area, or cold spots may occur which could lead to mould growth and to condensation and to corrosion of metal plate fasteners. It is also desirable to close the cavity wall at the wall head by insulating blocks. Figures 13.10 and 13.11 show bad and good practice in roof insulation at the eaves.

The occurrence of condensation in a pitched roof space can give rise to several problems:

- It can encourage the growth of non-woodrotting fungi such as Blue Stain mould. This is unsightly and makes the air in the roofspace unhealthy because of mould spore build-up.
- Mild steel conduits and fixings will corrode at a faster rate.
- It can cause leaching of soluble salts in clay tiles and timber sarking boards (Vegoda 1993).
- It reduces the thermal efficiency of exposed insulation in the roof space.

Positive ventilation of the roof space is usually essential and the form it should take will be dictated principally by the pitch of the roof. BS 5250 recommends that where the roof pitch is 15° or lower, ventilation openings at the eaves should be the equivalent of not less than 25 mm of continuous opening on each side. When the roof is of more than 15° of pitch this figure may be reduced to 10 mm. One way of doing this is to provide appropriate holes in the fascia and soffit boards. Ventilators at the ridge are not an acceptable alternative because they can cause pressure in the roof space to fall under certain wind conditions and this drop in pressure can draw moist air through the ceiling. It can be useful to close the main gaps in the ceiling below by weatherstripping around the access hatch cover, by sealing holes where pipes pass through the ceiling and by ensuring that ceiling roses are

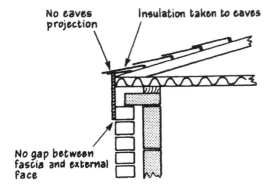

Figure 13.10 Poor practice: inadequate gap for air to enter behind fascia; insulation obstructs air movement; no eaves projection enhances risk of rain reaching top courses of brickwork if gutters become blocked or defective.

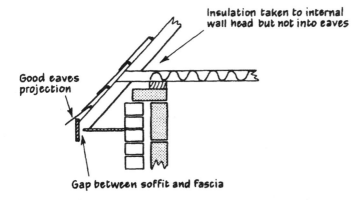

Figure 13.11 Good practice: gap for external air to enter behind fascia; insulation stopped short of the eaves; good eaves projection gives better protection to the external wall.

screwed tightly to the ceiling. Close covering of cold-water tanks is easily done and is useful. However, these measures will only reduce air flow into the roof space; they are unlikely wholly to prevent it. Ventilation of the roof space above insulated ceilings will always be necessary and is, indeed, a requirement of the Building Regulations.

As reported by Douglas (2006), however, the sealed roof using a proprietary breathable roofing felt such as Tyvek® "Supro" (with Tyvek® "SD2" air leakage barrier and vapour control layer on the warm side of construction) is a modern alternative for pitched roofs. It obviates the need to provide background ventilation in pitched roofs. The only caveat is that "in flat roofing a Tyvek® membrane may only be used as a protection layer over insulation or as a separating layer between metal sheeting and a supporting deck" (Dupont™ Tyvek® 2005).

Thus, these breathable polymer membranes are not without their problems. If exposed to sunlight for prolonged periods they are susceptible to ultra-violet degradation. Second, they can be degraded by chemical preservatives in plywood and other treated roof timbers if laid in direct contact with them. It is for these reasons that the manufacturer does not recommend the use of Tyvek® on plywood substrates.

13.3.3 Roof coverings

A common problem on slated or tiled roofs is known as "nail sickness". This occurs when the ferrous nail fixings become so corroded that the head of the nail no longer secures the slate/tile. As a result some or a few slates or tiles slip from their proper position. Loose and broken slates thus not only allow leaks into the building, they can also pose a safety hazard if they become dislodged from the roof slope and fall to the ground below.

Condensation on the back of the slate/tile or water penetration can trigger nail sickness (Hollis 2005). The use of proper underlay felt and copper (instead of galvanised mild steel), flat-headed roof nails can minimise the occurrence of this problem.

Tiled pitched roofs may suffer from frost damage but this is not common, and damage has been confined mainly to clay tiles used on roofs of shallow pitch. Tiles show signs of spalling on the top surface or may laminate. If the tendency to roofs of lower pitch continues, and if roof spaces become highly insulated, the risk of frost damage to both clay and concrete tiles cannot be ruled out. This is an area where further investigation is needed.

Sheeted industrial roofs are especially likely to suffer from copious condensation promoted by heat radiation to a clear night sky and particularly so on calm, frosty nights. This is an area which calls for expert design and the co-ordinated use of thermal insulation, vapour barriers and positive ventilation.

13.4 Summary

Roofs are highly vulnerable to wind and water problems. Their exposed nature means that they are constantly exposed to the elements. The risk of damage as well ordinary wear and tear therefore is fairly high. Detecting the precise source of a roof leak can be extremely difficult without the aid of sophisticated equipment such as a thermographic camera, as highlighted in Chapter 4.

Chapter 14

Services

OVERVIEW

The problems associated with building services are addressed in this chapter. It discusses the typical defects that can occur with installations and plant that service buildings.

14.1 Background

Failures in building services seem to lack the journalistic appeal of those affecting flat roofs. Probably, they are fewer in number, anyway, and are characterised more by a falling-off in performance than by total failure. However, some problems have been significant and relate either to heating or to plumbing and drainage.

14.2 Heating installations

14.2.1 Preamble

A great deal of the complaints made about heating installations have been of their cost in operation and the difficulties of using alternative fuels, through inflexibility in building design. Many housing estates, for example, are either wired for electricity or piped for gas but do not enjoy both, and many modern dwellings do not have chimneys and are not able to use solid fuel without alteration. These issues are essentially social and economic, and lie outside the scope of this book. In passing, however, it is worth restating that many occupiers have, as a consequence, chosen to use portable fuels, such as paraffin and liquefied petroleum gas, which generate large quantities of moisture during combustion and so increase the risk of condensation and the problems which stem from it.

There have been technical problems, however, in the post-war years, particularly those of scale formation in hard-water areas and corrosion.

14.2.2   Scale formation

Hot water pumped through radiators is, probably, the most common form of central heating. The composition of the water supplied is dependent upon the nature of the ground in the area in which the water originates and this will vary between individual water-supplying authorities. Rain absorbs carbon dioxide from the air to form carbonic acid and this, draining through calcium-bearing soils, will form calcium bicarbonate. When heated, this decomposes to calcium carbonate which is, for practical purposes, insoluble. Because calcium bicarbonate is so changed when water is boiled, it is said to cause "temporary" hardness. The equivalent amount of calcium carbonate in water is used as a classification of its hardness. Water draining through chalk may have more than 350 parts per million and is then classified as very hard. That derived from peaty areas is likely to have less than 50 ppm and is classified as soft. "Permanent" hardness is not removed by boiling and is caused mostly by the presence of calcium and magnesium sulphates. In a hot-water system, calcium carbonate will only continue to be deposited, and to build up to form a harmful scale, if the water is constantly replaced by further bicarbonate-bearing water. This will happen in a direct hot-water system in which the water heated by the boiler is drawn off and replenished by a new supply of cold water. Scale formed in this way in direct hot-water systems reduces the effectiveness of the boiler and, eventually, causes the entire blockage of water-ways. The rate of formation of scale increases with increasing temperature and with increasing degree of hardness. In recent years, many direct systems have been badly affected by blockage. Scale formation through permanent hardness can also cause blockage but the rate of formation is very slow. In part, failure has been increased by the introduction of smaller boilers with narrower water-ways.

Problems of blockage and furring of pipes, through scale formation, can be effectively prevented by adopting an indirect hot-water system. This is a system in which water heated by the boiler is not drawn off but is used to raise the temperature of water in an indirect storage cylinder. This it does by circulating through a heat exchanger within the cylinder so that it does not mix with the stored water. In consequence, and except for any small amounts of water lost by expansion or leakage, the same hot water circulates in the primary circuit and, after the initial deposition, no further scale can form.

Indirect systems are the most commonly used today and blockage by scale formation should be a declining problem.

14.2.3   Corrosion

Separate from the malfunctioning of central-heating systems through scale formation can be that due to corrosion. Severe problems have been encountered in indirect heating systems, where the same hot water is pumped continuously round a closed circuit. The heating circuits commonly involved have used cast-iron boilers, copper tubing and pressed steel radiators, and most problems have occurred in centralised systems heating a number of dwellings at a time. In an indirect system, the initial water quite quickly loses dissolved oxygen, and iron in the circuit then reacts with this water to form soluble iron products. This, in itself, is not particularly troublesome and an equilibrium is soon reached. Other factors, however, may cause the iron compounds to break down with the formation and release of hydrogen. Further iron compounds may then form and be broken down, until a sludge of less soluble iron compounds is formed. This and the entrapped hydrogen may together stop the circulation of water within the system and also prevent the proper functioning of valves. The causes of the generation of hydrogen and the breakdown of iron compounds into a less soluble form seem not to be understood fully, but copper, dissolved in small amounts in the water, is thought to be a contributory factor. When blockage has occurred, flushing out and chemical treatment to remove the sludge are the likely actions necessary but can be expensive.

Bimetallic corrosion, which is considered in some detail in Chapter 6, can be troublesome when oxygen is present in the water system but, with indirect systems, the oxygen content is generally low. However, it may be introduced through faulty design, for example if the circulating pump operates at too high a pressure and if air can enter the cold-feed pipe to the boiler. This can happen if the water level in the expansion or "header" tank ever falls below the level of that pipe. A common cause of pitting corrosion of pressed steel radiators has been the continual entry of air into the system promoting copper/steel bimetallic corrosion. Failures have also been associated with the formation of nitrogen or hydrogen sulphide by forms of bacteria. The frequent formation of air-locks, in addition to the initial air-locking which is not unusual in a new system, is an early sign of the likelihood of corrosion and blockage. If a check of the system shows no reason for leakage and entry of air, and the water level in the header tank is not too low and does not overflow either, then it is probable that a corrosion inhibitor containing a biocide, placed in the primary water supply to the boiler, will prove effective in reducing corrosion. However, it is essential that the local water authority is consulted before using an inhibitor because not all inhibitor/biocides are harmless to health, and in no circumstances must they be allowed to enter, or contaminate, the drawn-off water supply. They are for use only in closed primary circuits.

14.3 Chimneys and flues

Some years ago, extensive failures occurred in chimney stacks, particularly those serving slow-combustion solid-fuel boilers and closed stoves. Water vapour is always present in flue gases, which may also contain sulphur oxides and tar acids. As the flue gases pass up the chimney and reach the exposed part of the stack, the dew point is commonly reached and

Figure 14.1 Deterioration of brickwork typical of sulphate attack.

condensation takes place. The condensate migrates into the walls of the chimney stack, and may deposit tarry residues and, also, set up sulphate attack on the mortar, through the sulphur gases dissolved in it. Sometimes, additional sulphates contained in the bricks assist the attack (Figure 14.1).

Figure 14.2 Expansion of flue and the start of sulphate attack on the rendering.

As already described, sulphate attack causes the mortar to expand, and this leads to distortion of the stack, which can be extensive, particularly on rendered stacks (Figure 14.2). Stacks often bend in one direction and this may be due to preferential wetting of one side of the stack by rain predominantly from one direction. The condensate often contains salts, which absorb water vapour from the air and serve to keep affected parts damp. Both external brickwork and internal plaster on the chimney breast are likely to be discoloured by the tarry compounds, particularly at ceiling level: dampness will be visible in wet and humid weather. (If dampness, however, only corresponds with periods of rain, it is probable that the flashings and DPC in the stack are missing or faulty.) For some time now, the Building Regulations have required all new chimneys to be lined with materials suitably resistant to acids, and impermeable to liquids and vapours, and this problem should be one of the past, and of passing, rather than direct, interest. Repair to existing damaged chimneys will often require rebuilding of the stack and relining but, if distortion and damage are slight, it may prove possible to insert a suitable lining without dismantling.

14.4 Plumbing and drainage

14.4.1 Generally

The soils through which cold-water supply services pass can vary considerably, from comparatively non-aggressive light sand and chalk to heavy clays containing organic matter suitable for the development of bacteria which can assist corrosion. Some sulphate- and chloride-bearing soils can be very aggressive. Ground to a depth of up to 1200 mm is of main interest and here the natural soil may be contaminated, anyway, by a wide variety of builders' rubble. When corrosion or other faults do occur in buried pipework, repair tends to be costly. Protective tapes and wrappings are called for in many areas where galvanised steel is used. Protective tapes are usually recommended to be wrapped spirally around the pipe, with at least a 50 per cent overlap, giving a double layer of tape. Wrapping is a time-consuming task, however, and one which requires good supervision if gaps are not to be left which will present an area for concentrated attack. Some tapes, too, can be degraded by bacterial activity and fail unless they, in turn, are protected by an outer, inert plastics tape, also wrapped spirally to give a double thickness. Protective tapes have been damaged, and made ineffective, by insufficient care in back-filling the trench, in particular, by sharp-edged builders' rubble able to cut the tape. Trenches need to be back-filled, to a depth of around 300 mm, with selected fine material, such as sand, hand-packed around the pipe. Unfortunately, authoritative data on the best types of protective tape to use are lacking, partly due, no doubt, to the wide variety of corrosive and bacterial activity likely to be encountered. It may be useful to undertake a soil survey or to refer to existing surveys to determine the likely corrosivity and probability of bacterial activity. Lead is prohibited for all new work in England, Wales and Scotland. Copper may be used but this, too, can be readily attacked by rubble and will often need a protective coating. Polyethylene and unplasticised PVC are used also for cold-water supply and do not corrode, though they can be damaged by poor back-filling and by careless handling at low temperatures. They do not require taping, and unplasticised PVC pipes may well be a preferred solution in active soils. (Pipes made from many plastics are not suitable for hot water, and may soften and sag if placed near to hot pipes and where subject to undue temperature rises, such as in airing-cupboards.)

14.4.2 Polybutylene pipes

The solution to avoiding the problems referred to above concerning scale build up and corrosion in pipes is to use a modern plumbing material such as polybutylene for the pipework. For example, Hepworth's Hep$_2$O polybutylene hot and cold plumbing pipework offers the following advantages over copper:

- No scale build-up
- Corrosion free
- Quieter
- Solder free – push-fit joints
- Cool to touch
- No burst pipes.

Under damp conditions, pipes passing into the structure through brick, concrete and plaster can be corroded. Pitting corrosion of stainless steel can occur when the unsheathed steel passes through breeze blocks, possibly caused by the presence of small amounts of chlorides. All metal pipes passing through these building materials should be suitably coated, taped or sheathed: it is safest to assume that, sometime or other, they will become damp. However, failures are few and the problem is not a significant one nationally.

Internally, corrosion in plumbing services is caused principally by the use of mixed metals in the system. Of the mixed metals used, the combination which has given rise to most failures has been copper/galvanised steel. It is the galvanised steel which will corrode. The corrosion rate increases if the water contains much oxygen or chloride ions. The chief risk is when water flow is from the copper part, of the system to the galvanised steel part and when new copper components are introduced into an old system using a galvanised steel tank. However, if flow is from a galvanised steel tank to the copper, and if copper in solution in the hot-water part of the system does not find its way back to the galvanised tank, for example via the expansion pipe, the risk of corrosion is greatly reduced. The use of galvanised steel and copper together in a hot-water system will almost certainly lead to rapid failure of the former. Such a combination is unlikely in new buildings but many older ones contain galvanised steel pipes used before copper came into favour. The introduction of a new copper cylinder in such cases can cause rapid failure. Stainless steel is now moving slowly into use but acceptance is retarded by difficulties in jointing, caused, in part, by its low thermal conductivity which requires different techniques of soldering. The fluxes commonly used for soldering copper pipes are based on zinc chloride and were used with stainless steel when the latter was first introduced. Severe pitting corrosion has occurred when such fluxes have been used excessively and when a long time has elapsed between making the joint and passing water through the pipe. Chloride residues need to be completely removed from stainless steel after application, otherwise pitting corrosion occurs. The need for such care, and the evidence of severe corrosion in its absence, has retarded acceptance of stainless steel. Fluxes based on phosphoric acid have been introduced, which are believed to be much safer. In the absence of chlorides, stainless steel in plumbing systems should not, itself, corrode and will not cause corrosion of copper.

Minor troubles can affect ball valves though this has been of small significance. Ball valves can stick in the open position, leading to overflowing or flooding if the overflow pipe is blocked or defective. Generally, this is merely a sign of the need for a new washer but the seating may be eroded by high-velocity water discharge. Many older floats, formed of two copper hemispheres soldered together, can corrode, become perforated and partly fill with water, which then prevents the closure of the ball valve. They may drop away from the float arm, leading to rapid overflow. Mostly, however, plastics floats are now used and these do not corrode and give much better long-term performance. Calcium carbonate deposits in hard water areas forming on the piston of the valve cause this to stick and are a common reason for overflowing.

Finally, a few words on frost damage. Pipes placed above insulation in a roof space will get cold. Lagging will slow down the rate of heat loss but, if heat gain into the roof space either from below or from other pipes is too small, freezing may still occur. Even if pre-formed polyethylene lagging is used it will merely postpone eventual freezing of "live" water pipework in an unheated building. With the likelihood of increased insulation in roof spaces, the risk of freezing may also increase, unless the pipes are taken under the insulation wherever possible, or some other steps are taken to ensure a low heat input to the pipes.

14.5 Electricity supply

All electric cables give off heat in use and this is usually dissipated without any difficulty. However, cables can get overheated if placed beneath loft insulation, behind insulated dry lining or if placed in a position where the temperature of the surrounding environment is high. The insulation of the cable will then be damaged and there can be a risk of a short-circuit and, possibly, fire. Cable in power circuits is more at risk than cable in lighting circuits, for the former is more likely to be loaded near to full capacity. Electric cable should not be covered by thermal insulation nor, ideally, should it be used where ambient temperature regularly exceeds 30 °C. Where this is not feasible, cables will need to be de-rated in accordance with the regulations prepared by the Institution of Electrical Engineers. It may be worth noting that a cable de-rating factor as low as 0–5 will be needed where cable is insulated on both sides. There can also be an interaction between PVC cables and expanded polystyrene often used for insulation and this can cause degradation of the PVC. The two should be kept apart.

As noted by Noy and Douglas (2005), loose connections can trigger an electrical fire as well as cause malfunction of the electrical system. In addition, an inadequate cable rating to appliances such as shower units and cookers can also cause heat build-up, thus increasing the fire risk.

14.6 Summary

Services account for at least a third of the capital cost of a building (Duffy 1993). In some cases, in highly serviced buildings, they can form nearly half of its initial cost. Wear and tear of services is greater than for most other parts of a building. This explains why the useful life of services is much shorter than that of other components.

Chapter 15

Failure patterns and control

OVERVIEW

This chapter addresses the appraisal of failure patterns and how they can be controlled. It attempts to formulate a strategy for reducing problems, particularly in new construction.

15.1 Background

This final chapter is different in nature from those preceding, and, after commenting on failure patterns, suggests reasons why avoidable building failures occur. Such reasons involve consideration of the structure of the industry, the particular problems associated with innovation, and control systems. These essentially "software" aspects of the avoidance of building failures merit as much, and probably more, attention than the "hardware" aspects dealt with already. It is, indeed, probable that the greatest scope for improvement in building performance lies more with the former than the latter.

Every type of building contains defects to some degree. The consequences of such defects may be minor but others are more important and may affect the appearance, value and usage of the building. However, the cost of rectifying or (worse still) not rectifying building defects could be extensive. In more serious instances, they may pose a hazard to health and safety.

While there is a considerable consensus of opinion on the technical reasons for failure which, it is believed, has been reflected in the preceding chapters, the "software" aspects are more speculative. The views on these are many and varied. It is hoped that those expressed here will be of some interest and possible value.

15.2 Ongoing defects

15.2.1 New buildings

Why is it, despite our increasing understanding of construction technology and materials, that building failures still occur? The reasons are not hard to find. Modern buildings are more sophisticated, as are their systems and facilities. Many of these advances make the chances of mistakes occurring in the construction process greater. The use of innovative materials and new construction techniques may not be fully tried and tested for every situation. This can also result in unforeseen problems immediately after a building's commissioning as well as during its occupancy.

Fast track building and shorter construction times are becoming increasingly common exigencies for the industry. All of these changes have put enormous pressures on designers and contractors to produce defect-free buildings that meet the needs of property users.

15.2.2 Existing buildings

Recurrence of building defects is often triggered by acts of omission. Neglect and poor quality maintenance are major influences on the incidence of important defects such as dry rot, for example (Palfreyman and Urquhart 2002). Moreover, failure to repair properly or on time can also activate further defects (Wood 2003).

Buildings that are partially or totally vacant are more vulnerable to dilapidation and defects. In particular, empty dwellings pose a direct risk of damage to adjoining properties through dampness and other infestations. For example, around 0.75 million dwellings in the United Kingdom are classed as empty – or 3.4 per cent of the housing stock (ODPM 2003). According to the ODPM (2003), "the longer a property is empty, the greater the likelihood that its physical condition will deteriorate, making it more difficult to bring back into occupation at a later date and increasing the likelihood that it will impact on neighbouring properties".

Acts of commission comprise a range of factors that can also contribute to the ongoing nature of building defects. Poor quality of original construction, of course, potentially has a devastating legacy on the aftercare of a building (Beukel van den 1991). This can be exacerbated by inappropriate repairs or defective/ill-considered adaptations (Douglas 2006). Climate change, resulting in increased flooding or precipitation, may lead to unforeseen water ingress (see Graves and Phillipson 2000). Vandalism and other forms of abuse can make all of these problems worse.

Massari and Massari (1993), for example, showed where misguided application of waterproof lining repairs to a building can cause the rise of moisture in a wall. If the wall is lined on both sides, this can lead to wicking of moisture up the masonry.

15.3   Review of causes

In the various studies by the Building Research Establishment's (BRE) Advisory Service, the causes of failures were also analysed to indicate whether they were due to faulty design, poor execution, the use of poor materials or through unexpected user requirements (BRE Digest 268). Faulty design was taken to include all cases where the failure could reasonably be attributed to a failure to follow established design criteria: 58 per cent of all failures were found to lie in this category. Faulty execution, defects attributed to a failure of the builder to carry out a design satisfactory in itself and properly specified, accounted for 35 per cent. Only in 12 per cent of cases did the materials or components fail to meet their generally accepted performance levels. Some 11 per cent of failures were caused by the users expecting more from the design than the designer anticipated. (There was some overlap between these categories.) A more recent analysis within the defect reporting systems of the PSA already referred to showed that 51 per cent of the causes of defects were due primarily to shortcomings in design, 28 per cent to inadequate workmanship and the remainder were due to other causes including the failure of materials (Construction 1983). A BRE study into faults in conventional two-storey house building in England found that 50 per cent were attributable to design, 41 per cent to site and only 8 per cent to other causes including materials (IP 15/90). The fact that far more failures occurred through inadequacies of design or execution than through faulty materials was a result which is confirmed by most subjective assessments.

Figure 15.1 summarises the main sources of defects in the United Kingdom. For a more detailed analysis of statistical data relating to building defects in this part of the developed world, however, readers should consult the excellent BRE series on building elements (Harrison 1996; Pye and Harrison 1997; Harrison and de Vekey 1998; Harrison and Trotman 2000;

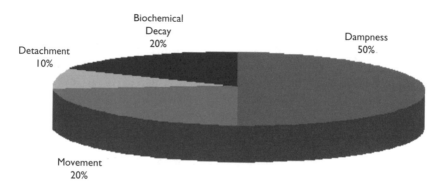

Figure 15.1 Basic analysis of defects.

Harrison and Trotman 2002). Houghton-Evans (2005) provides a useful summary of these and other similar studies.

15.4   Recap on reasons for failure

There seems little doubt that a major reason for failure in construction is the complex structure of the building industry. Despite the calls for closer integration of design and construction, made many times over the years, these two vital roles are still essentially separate. The manufacture of the basic materials and components, too, upon which the industry depends, is dispersed among many firms. These vary in size and type, from large firms, often with their roots in other industries, for example the chemical industry, to small firms supplying minor components. The building contractors, in the main, are small, with most employing fewer than ten operatives. In the United Kingdom, there are more than 30,000 general builders and a similar number of specialist contractors, such as painters and plumbers. Some 70 per cent of total output is attributable to contractors employing fewer than 20 people. In recent years, there has been a marked change to extensive subcontracting by the main contractor and a noticeable trend towards labour-only contracting. These structural complexities in this large and diffuse industry are conducive to difficulties in management, in education and in training; in the communication and application of technical information; and in the understanding and acceptance of responsibility. Construction, as a whole, has moved away from being craft-based towards being industrialised. However, it is currently in a difficult in-between state, and the clear loss of craft skills has not been replaced by the type of relatively fault-free product typical of truly industrialised processes. The loss of craft skills has been matched at a more professional level by a reduction in the number of those able to comment sensibly on the likely interactive effects of changes in materials, components and procedures. Many of the problems which have beset the so-called industrialised building systems have stemmed directly from the general inability in the construction industry to understand these interactions, together with inadequate matching of site skills to the new technology (Warszawski 1999).

It is difficult, clearly, for technical information at the right level to "flow steadily and smoothly" to the right recipients in such a diverse and changing industry. Shorter lines of communication, which obtain in more coherent and concentrated industries, would do much to improve the flow but there are few signs, as yet, of change in the basic structure of the industry. In some cases, the technical recommendations, although available, are not readily available and while they do state "the truth and nothing but the truth" it is, sometimes, not "the whole truth". Thus, some failures, though probably a relatively small portion of the whole, do stem from knowledge which is less than comprehensive. It is probable that most failures, however, could have been prevented by the proper application of existing technical knowledge. Technical information is produced in many forms but its diversity equals, rather than matches, that of the industry. The knowledge within the United Kingdom is probably as comprehensive as that in other countries but it is fragmented. Reliable, self-contained and comprehensive guides to specific design and building operations, written at the various levels needed to match the skills and abilities of the constituent parts of the construction industry, have in general not been readily available in the past.

The setting up at the BRE in 1981 by the Minister of Housing and Construction (at the then Department of the Environment) of a Housing Defects Prevention Unit seemed an encouraging sign. This Unit sent to contact points within local authorities short, sharp advice through its well-illustrated Defect Action Sheets aimed to prevent, in housing, the most common defects in design and in site practice. As housing, both in new construction and in repair and maintenance, accounts for about one half of total construction output, this initiative seemed timely.

However, political changes since then resulted in the formation of new central government departments taking over responsibility for various construction matters. For example, the Department for Communities and Local Government is responsible for the Building Regulations in England and Wales. It also deals with the Decent Homes Initiative, which aims to bring all social housing into decent condition by 2010 (Douglas 2006). The Department for Trade & Industry (DTI) looks after construction industry matters, mainly its business and productivity issues. It is in charge of the Constructing Excellence programme, which "aims to deliver improved industry performance resulting in a demonstrably better built environment".

In 1997, the BRE was effectively privatised. It is now administered by the BRE Trust, a registered charity. Despite its change of status the BRE is still providing feedback on building problems in the United Kingdom. It continues to provide world-class independent advice and information on building performance, sustainability, innovation and fire safety.

In addition to the problems engendered by the structure of the industry and the difficulties it causes for the capture, dissemination and implementation of technical knowledge are those of innovation, control and product certification. The construction industry has to operate within a complex framework of control, which is not uniform within the United Kingdom and within which, in recent years, the final responsibility for failure seems unclear. Moreover, the post-war years have seen rapid innovation in both products and methods. While innovatory pressures have been considerable, the requirement for adequate control and certification of these has lacked "bite". Too often within the innovatory decision-making process the views of those with a mature knowledge of building performance have carried less weight than those pursuing technical change.

15.5   Problems of innovation

Pressure for innovation may arise through events affecting the nation as a whole – for example, the present need to conserve energy and the consequent legislation which has exerted a pressure for new or improved ways of insulating buildings. Actual or potential scarcity of resources, too, has been a potent influence on innovation. The resources may be of labour and particular skills, or of materials. A related pressure may be the need, assumed or real, for speed. Such pressure usually follows situations of greatest social and economic change, for example wars, and was the main one behind the non-traditional construction methods urged after both the First and the Second World Wars, with unhappy results. The detrimental environmental effects of accumulation of waste of other industries, and the costs of its disposal, have also resulted in pressure on the building industry, because of its large market size, to take part of the waste as a raw material. Inflation, too, tends to cause pressure for labour-saving innovation in industries with a high labour content, such as the building industry. These strategic pressures find expression in entrepreneurial innovation where, rightly or wrongly, it is seen as likely to increase profitability or as essential to maintain competitiveness. Quite often, the creative thought and its transformation to a marketed product or method derives from other industries. The chemical industry, and, in particular, the plastics part of it, is the most obvious example.

When the general, social and economic system remains stable, the pressure for innovation is reduced: where it is unstable, innovation is a natural reaction. Change is not necessarily progress. It may create healthy competition but may also destroy skills which cannot so readily be re-created. Poor innovation can, unfortunately, be self-perpetuating, creating the need for remedial innovation. In contrast, the more radical and better the innovation, the more it helps to restrict the climate for further innovation. It would be interesting to determine, if such a thing were feasible, the relationship in the building industry between numbers of innovations and general quality of building: it might well be inversely proportional. These pressures do not arise from consumer choice but from industry, commerce and Government. This is, of course, not an unusual state of affairs in other sectors of industry, and stems in part from the lack of a suitable mechanism to obtain users' requirements in any meaningful way and the inability of many users, through ignorance, to express valid opinions. While these pressures for innovation are real and active, there are constraints to responsible innovation, several of them peculiar to the building industry. The major constraint lies in its organisation. The building industry is not an entity which, by its organisational and financial structure, permits the injection of large resources to encourage responsible innovation and there seem few signs of early change. Innovation, in fact, usually comes from the manufacturer and supplier of materials and components.

Innovation in the building industry is a somewhat random process, most often directed at reducing initial cost, though protective pricing, to provide a margin against uncertainty, may make this aim imperfectly realised. The lowest initial cost concept, a corner-stone of most of the associated advertising, almost always results in a reduction in the quality of materials, services and structural components, or in the amount of usable space in the building. Such innovation slowly tends towards shortening total life and increasing the need for maintenance. This is in opposition to the true needs of conserving the building stock and reducing the need for repair, which is higher in labour content than that for initial building. It is seldom that innovation takes place against a criterion of total costs in use. To be successful commercially, the time scale for innovation, development, production and marketing needs to be as short as possible but this contrasts with the basic development in building which has evolved over a long period. The tendency is to innovate at the trivial end, rather than to consider control of the whole system and its likely future needs. The innovator seldom has a clear grasp of the required future state of the building stock—only a knowledge, usually, of one isolated facet of the state he aims to change. The most common danger is to ignore, or to be unaware of, the often complex interactions which the innovation will alter. The innovation is nearly always concerned with providing an alternative to the primary function of an existing material or system but seldom has regard to important, though often ill-defined, secondary features. A typical paradox, through this lack of total understanding, is the way improvements in the prevention of heat loss may increase the risk of condensation or rain penetration: the solution provided by the innovation is often subtly different from that intended. A further danger is the inability to appreciate, or at least to take account of, erection accuracies and the level of supervision reasonably possible on building sites, and failure to provide margins against slight misuse, in both design and construction.

In the longer term, and at its most extreme, there is a possibility that innovation will make existing knowledge and skills redundant. If the innovation is enduring and valuable, this may be acceptable but, if not, the nation will be the loser. But, perhaps, the biggest danger of all lies in the inadequate mechanism for control of the innovation to safeguard the user.

15.6   Minimising building failures

15.6.1   Means

Given the problems of diversity of participants and techniques used, uniqueness of every building project, outside working conditions, and so on, it is not too difficult to appreciate why failures in construction are more prevalent than those in other industries. Indeed it is the very nature of the building industry itself, which makes it difficult to eliminate such problems entirely. Moreover, some building professionals, such as Puller-Strecker (1990), have argued that building defects will never be entirely eliminated:

> If by "building defects" we mean causes of disappointment in the initial or long-term performance of buildings, then building defects, like the poor, will always be with us, and for the same reason. The poor will always be with us because there will always be those who are less well-off than the rich, or even the average or are simply less well-off than they feel they have a right to expect: building defects will always be with us because there will always be aspects of building performance that are less good than the best, or even average, or less good than we feel we have a right to expect.

That statement is more pragmatic than fatalistic or pessimistic. Nevertheless there are a number of measures that would go some way towards minimising if not avoiding common construction failures:

- An improvement in the "lowest tender" system, without generating excessive inflation of contract prices, should be sought (Houghton-Evans 2005). Cheapest price does not necessarily mean best value for money, particularly in the long term. It is for this reason that some clients (e.g. Ministry of Defence) use a best value for money approach. This takes into account quality of product as well as lowest whole life cost.
- Increased/improved feedback from builders, research bodies, maintenance engineers/surveyors, and users to designers/specifiers and better feed-forward from designers/specifiers to maintenance managers and users. Increasing the accessibility to information for building professionals, managers and technicians will go a long way to achieving this goal.
- Regular on-the-job training programmes for operatives and site staff.
- Improvements could be made in the quality of graphical and written communications. For example, three-dimensional drawings could be used more and greater use could be made of national specifications. In addition, training could be given to site staff on improving verbal communication skills.
- Better and more available guidance on commissioning buildings and their services (see Odom and Debose 1999 below) and on defects avoidance (see next section).

15.6.2 Guidance on building defects

The literature offering advice on building failures and inspections is expanding every year (e.g. Hinks and Cook 1997; McDonald 2002). Typical sources giving sound guidance on general defects are National Building Agency (1983), Cook and Hinks (1992), Son and Yuen (1993), Richardson (2001), Carillion Services (2001), Marshall *et al.* (2003), and Noy and Douglas (2005).

The BRE, in particular, publishes a very informative collection of literature on a wide range of building problems and solutions (see list in the Bibliography) from alkali aggregate reaction in concrete (Digest 330) to wall tie replacement (Digest 329). In addition, its Good Building Guides (GBG) and Good Repair Guides (GRG) especially are very useful. GRGs offer "concise guidance on the principles and practicalities for achieving good quality building" (e.g. see GBG 21). GRGs, on the other hand, "are accessible, illustrated guides to defect diagnosis, assessment and repair" (e.g. see GRG 12).

The texts in the Bibliography comprise a reasonably comprehensive list of works on Building Pathology and inspection-related subjects. In particular, the books by Watt (1999) and Harris (2001) offer informative perspectives on this discipline from the United Kingdom and United States respectively. The works by Addleson and Rice (1992) and Cook and Hinks (1992), for example, give useful tips on the methodology for investigating building defects. Haverstock (1999) and Hollis (2005) respectively provide excellent guidance on building design and property survey issues. For detailed guidance on the lessons learned from forensic investigations into design and construction failures, see Carper (1989), Croker (1990) and Kaminetzky (1992).

Probably the most significant guidance in recent years dealing with defect avoidance in the United Kingdom, however, is derived from the work of Housing and Property Mutual (HAPM). Their defects avoidance manual (HAPM Ltd 1991) was followed by their guide to defects avoidance (HAPM Ltd 2001). These much needed publications are aimed at reducing the incidence of failure caused by design and construction shortcomings.

According to HAPM Ltd (2001),

> Many who approach the subject of building failure do so without recognising that defective design often stems from not understanding the difference between a defect and its consequences and that, conversely, the consequences of a defect cannot be "made good" until the defect itself has been diagnosed and remedied. It is crucial that the difference between cause and effect that the "HAPM Guide to Defects Avoidance" seeks to address.

15.6.3   Cost and types of defects

There are no precise data on the total national cost of repairing avoidable defects. Only in a few major organisations in the public sector are there systems in operation which aim to record defects systematically. The private sector is too diffuse for any worthwhile attempt to have yet been made to collect such data. In the housing side of the private sector, many defects will be rectified, anyway, by uncosted "do-it-yourself" activity. Typical maintenance budgets in both public and private sectors cover normal maintenance due to fair wear and tear and the cost of cleaning, as well as repair. In considering costs, problems of definition arise in deciding what is an avoidable defect and whether the estimate of cost should include not only that incurred in rectifying, physically, the defect, but also the consequential costs, such as loss of rent or production and temporary resettlement. Consequential costs may be much greater than direct repair costs and the wide range of guesses at the cost of failures is doubtless, partly due to the different boundaries within which the costing assessment is made. Nevertheless, the defect-reporting systems of the now defunct PSA revealed that more than 10 per cent of the Agency's maintenance bill was attributable to preventable defects occurring within about 5 years of construction (Construction 1981). Nationally, the total cost of repair and maintenance is estimated to be in excess of £45 billion (Douglas 2006). The implication is that preventable defects are likely to be costing somewhere in the region of £4.5 billion a year.

Reliable data on the relative proportions, nationally, of different types of defect by number and significance are not easy to come by. Since the early 1970s, however, analysis has been made of over 500 investigations undertaken by the BRE's Advisory Service. The reports emphasised that the problems were not necessarily representative of the generality of building defects. In the analysis, dampness, whether from rain penetration, condensation, entrapped water or rising damp, accounted for half of all the defects investigated. Condensation was the largest single cause of complaint in council housing and rain penetration in private housing. It is probable that, since the analysis was made, condensation has retained its pre-eminence as a problem in council housing. Condensation, of course, is not necessarily caused by a design or construction failure, though it may be exacerbated by them. Rain penetration, on the other hand, clearly is, and the analysis showed that most cases reported were those of penetration through walls including parapet walls, followed closely by penetration through roofs. Cracking, both structural and superficial, and detachment of finishes, such as rendering, brick slips and tiles, accounted for some two-thirds of all other reported defects. The cost significance of the various defects analysed was not reported.

As we saw in Chapters 2 and 3, the analysis of defects starts, generally, with a classification of symptoms, for example a damp ceiling as a symptom.

This is the easy part. A more penetrating diagnosis will then be required to determine the cause of the symptom, for example split flat roofing felt. Still further analysis will be needed to determine the cause of the splitting, such as differential movement between felt and substrate. The ultimate cause of the differential movement will lie with one or more of the agencies mentioned in Chapter 4. These agencies are the real physical, chemical and biological causes of failures. The deeper and more efficient the analysis, the greater will be the success of a remedial measure. All too often, the analysis stops prematurely and a facile solution is adopted, which may be unsuccessful or only partially successful. Good design and construction, aimed at preventing premature failure, should start with an understanding of the basic agencies likely to affect the element of structure in question, assess their effects and then design to minimize or prevent them. In general, this synthesis is performed intuitively rather than objectively.

15.7   Towards better control

15.7.1   Background

Any change in the basic structure of the building industry is likely to be slow and brought about by economic forces rather than by any conscious desire to reduce the incidence of failures. It seems probable that the continuing fall in the number of building operatives, and the increasing cost of the labour, will comprise an economic force likely to enhance the development of semi-industrialised building techniques, leading to faster erection and requiring less skill. In the long term, this should improve the interface between design and construction, by the encouragement of "design-and-construct" organisations. The harder economic climate is also likely to reduce the number of small contractors and lead to amalgamations which will, in theory, provide better targets for technical information and for feedback.

The principal organisations, both public and private, have the ability to reduce the risk of failures by taking formal steps to provide an effective interface between design and maintenance. Maintenance organisations are the repository of much knowledge of the nature of failures and their cause, and such knowledge fed back into design would be of great benefit. A dialogue at the design stage could do much to make designers more conscious of the implications of their design decisions and the ability of contractors to fulfil them safely, and should lead, if such a dialogue is conducted with tact, to alternative and less-sensitive design decisions. The large organisations might have the resources to adopt the formal collection and feedback of information relating to failures, but even the more modestly sized organisations could, it is suggested, establish such a dialogue. The PSA design/maintenance liaison (DML) system was set up as a vital link to improve the quality of building and to eliminate premature maintenance costs. It had clearly proved its worth within the public sector. But since the Agency's demise in the late 1980s such a system was discontinued.

There is a general need throughout the construction industry to discourage designers from using novel solutions to problems where standard solutions are already available. There is still a need to improve the control of innovation. The British Board of Agrément's function is to assess, test and, where appropriate, issue certificates in respect of materials, products, systems and techniques used in the construction industry, particularly those of an innovative nature, in order to facilitate their ready acceptance and their safe and effective use. Agrément certificates provide the Board's opinion of the fitness for specified purposes of materials, products, systems and techniques taking into account the context in which they are to be used. While a large number of certificates have been issued there is no compulsion for an innovation to be referred to the BBA. The process of designing and building is often achieved with little involvement of the professions within the industry, and as a result products and processes may be used which are unsuitable. If BBA certificates were demanded, if only for certain categories of materials and components, then manufacturers would find it more worthwhile to pay for the evaluative testing and to modify the innovation if necessary. As it is, a manufacturer accepts the need to obtain a certificate only if he sees this as of commercial benefit. Building Regulations provide some brake on innovation. They are, however, made for specific purposes of health and safety, energy conservation and the welfare and convenience of disabled people. They are not concerned with other aspects of performance such as serviceability.

Innovations should be examined in relation to total-life costs, with particular emphasis on improving durability and reducing the need for maintenance. There is a need to move away from the current concept of trying to encourage innovation by paring safety margins and quality, and to move towards innovation aimed at reducing obsolescence and providing greater adaptability. The risks of innovation should be more truly balanced against the claims by indepth studies of the whole system. As a start, priority should be given to examination of innovations affecting the structure of buildings. A second order of priority should attach to those innovations which, while not directly affecting safety, may, if they fail, cause big expenditure. Studies of innovation by the innovator, and any control system, should include

- full consideration of the life aimed at; the positive evidence for the likelihood of its achievement;
- the likely extreme and normal environmental conditions;
- the probable loads; the factors which might cause failure;
- the probability of occurrence of such factors;
- the consequence of failure;

- the warnings needed for the designer and builder;
- the quality of control desirable; the strength of control which can be exercised by the innovator over the use of the product;
- the skills needed in construction likely to be available; and,
- the margin of permissible error and the type, and ease of undertaking, of any future maintenance.

In support of responsible innovation, there is a need for greater knowledge of how buildings and their elements actually behave, as opposed to the assumptions so often made and enshrined in British Standards. This implies major planned and purposeful site surveys. Associated with these should be an accelerated development of performance requirements and their related evaluation procedures. In recent years, there have been significant developments in the field of quality assurance, in particular the Government initiatives set out in a White Paper and BS 5750 "Quality Systems", the latter being concerned with the quality of entire management systems. There are now BSI Registered Firms – firms certified by BSI as having a quality management system in accordance with BS 5750 (ISO 9000). Some of these firms are also holders of BBA certificates. Failures are not often due to the poor quality of materials and components. Nevertheless, the appraisal and control of the quality of these is more marked here than it is in design offices and on sites where it is most needed. Doubtless, this is because it is easier to do so. However, steps are needed to improve significantly the quality control of designs and specifications. In design offices, much of the detailing and specification is the task of the more-junior and less-experienced designers, who have little knowledge of what is practicable on site, and the abuses to which materials and components are subjected. More effort is needed to develop standard solutions and details. In general, the typical site operative will do what is quickest and easiest. If details and whole designs can be devised so that what is quickest and easiest is also sound and durable, then a big advance would be made. Design would benefit considerably if design offices were organized to ensure that the experience and knowledge of senior designers is brought to bear, both in this vital area and in the independent checking of detail.

15.7.2 Commissioning buildings

Building failures seem prevalent throughout the globe. According to Odom and Dubose (1999 and 2002) in the United States, for example,

> The building industry seems baffled about the prevalence of building failures. Many wonder: Why isn't the rate of building failures declining?

Why, in spite of better technology, increased training, and more sophisticated building systems, does the industry seem unable to prevent the next failure?

It's not from indifference, and it's not that we can't prevent buildings from failing. We know we can, because we can fix them once they do fail. The reason we aren't coming to grips with this issue is simple: the right people, the people most entrusted with how our buildings perform, are not receiving feedback on their building's performance. No report cards are available to tell us if we really did a good job. We don't know why our last five buildings apparently worked well but the next one didn't, despite being apparently designed the same way. Metrics may say that we did a good job, yet clients keep complaining about building failure and the construction litigation business keeps growing. In short, institutional knowledge is not increasing because the performance feedback mechanism is so poor.

Buildings fail because of widespread misinformation about building failure itself. We believe if our last five clients didn't complain, then surely we gave them a good product. The frightful reality is that some of our buildings are at risk and we don't even know it. Moreover, many of us don't know what puts those buildings at risk, or how to minimize that risk cheaply and effectively.

New building failure is a function of unrealistic owner expectations, complexity of the design and construction process, and contradiction among project drivers (cost, schedule, and quality).

We have observed that most buildings fail because of flawed concepts, not incorrect implementation of the design details. The following three elements summarize the Walt Disney World and CH2M HILL model and are integral to its success in improving the performance of new buildings:

1. Establish specific written design and construction guidelines at the project's inception, and distribute them to all relevant parties.
2. Use periodic peer reviews throughout the design and construction process to compare results against the original design and construction guidelines.
3. Implement proper start-up techniques for complex installations such as the HVAC system to verify that the building is operating correctly before it is occupied.

In other words, what is needed is proper commissioning of buildings. Commissioning a newly completed building is essential to determine whether or not its performance matches design and user expectations. Commissioning is defined by ASHRAE (2005) as: "the process of ensuring that systems are designed, installed, functionally tested, and capable of being operated and maintained to perform in conformity with the design intent". The guideline states that "commissioning begins with planning and includes design, construction, start-up, acceptance and training, and can be applied throughout the life of the building". Commissioning ends with assuring that the operators are trained and operation and maintenance (O&M) manuals are available and accurate. (See http://www.peci.org/library.htm website for informative resources on commissioning buildings.)

15.8   Summary

Achieving a property of good quality starts right at the beginning of the development process. Thus, even at the inception stage consideration must be given as to how to ensure that the completed building has minimal defects. Figure 15.2 illustrates the key interventions required to help achieve this goal.

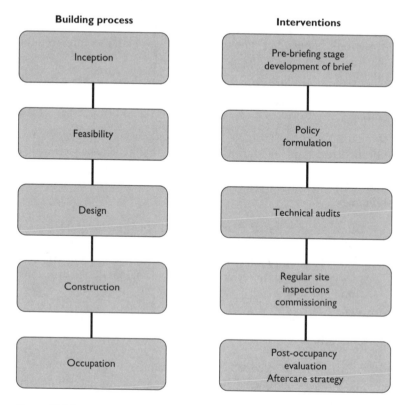

Figure 15.2 Interventions for achieving quality in construction.

Furthermore, the need for improved quality of site operations is clear but the way forward is obscure. Sooner or later, the main contractor will need to regain control over subcontract labour, and site supervision will need to be more concerned with the quality of the finished product and less with accepting shoddy building in the interests of more harmonious site relations. This control may, in part, be brought about by financial penalty or by some return to directly employed tradesmen. It may also be assisted, particularly in housing, by the development of industrialised systems which require less skill and less labour for erection. Any such systems will need to be designed with much greater care than in the past. Ideally, the erection of the individual components should only be possible in one way, and that one should be the easiest. Wherever possible the design of any building should facilitate ease of construction. Improvements are needed in the dissemination of technical information at all levels, but particularly between the designer and the contractor. Design information should be complete and the builder should not need to spend his time in ascertaining the designer's real intentions or in devising his own, often inadequate, design solutions to problems. Drawings and specifications should be complete and the information on, or in, them easily found by using some structured system. As already mentioned, there is not so much a lack of information as a poor match between the level of information and the recipient. There is a particular need for more detailed, specific, self-contained and readable guidance to be given on the design of individual building elements, and for parallel guidance to builders and operatives on site to meet the design intentions. British Standards could provide the starting-point for such guides but are not, as they stand, appropriate.

There is also a need for designers to be aware, in a far more specific way, of the likelihood and cause of failures. Their education and training needs to encompass the most common types of failure and their cause. Designers also need, during their training, guidance to the sources of reliable information on design principles and solutions, to avoid failures. It is hoped that this book will go some small way in assisting such education and training.

Glossary

abrasion   The wear or removal of the surface of a solid material as a result of relative movement of other solid bodies in contact with it (Addleson and Rice 1992).

absorption   The entry of a fluid into a solid by virtue of the porosity of the latter (CIB 1993).

absolute humidity   This is a measure of the actual quantity of water vapour present in the air at any given time. It is a variable quantity measured in kg/m^3, and can best be found by experiment using a glass jar and a "U" tube containing calcium chloride (Oliver *et al.* 1997).

acids   Compounds of acidic radical with hydrogen which can be replaced by a metal (usually sodium) either wholly or in part. Acids turn blue litmus red (Addleson and Rice 1992).

additive   Strictly speaking, an additive is a material that is added as part of the cement or other binder during its manufacture. Examples of such additives would be cement process grinding aids such as polyethylene glycol. The term, however, is also used to describe ingredients such as plasticisers, waterproofers and other special agents, which are pre-mixed with the cement or plaster prior to it being bagged (Oliver *et al.* 1997).

admixture   A special material added in small quantities to cement or concrete mixes which alters the properties or strength of the product (Oliver *et al.* 1997). Examples of admixtures are air-entraining agents, waterproofing compounds which are added during the mixing process on site. This is the essential difference between admixtures (added on site) and additives (added during manufacture).

adsorption   The attachment of a substance to the surface of a solid by virtue of forces arising from molecular attraction (CIB 1993).

aetiology   The study of causes of things.

ageing   Degradation due to long-term influence of agents related to use (ISO 15686-1:2000).

air dry   The condition of a material in an ordinary indoor, inhabited environment with a relative humidity not exceeding about 70 per cent (Oliver *et al.* 1997).

air drying (natural seasoning) Converted green timber can be dried to a moisture content of about 16–18 per cent by stacking the boards in such a manner that a free flow of air is possible over all surfaces and the stack is protected from the rain. It can be a time-consuming process, and with some timbers the air drying rate may be as slow as 25 mm (1 inch) thickness of timber drying per year.

Many damp building materials may dry out as a result of being exposed to the air in a building. If they are protected from dampness, they will achieve a moisture content in balance with the prevailing ambient humidity conditions (Oliver *et al.* 1997).

alkali Soluble bases or hydroxides. A base is a compound which reacts with an acid to yield a salt and water only. Alkalis turn red litmus paper blue (Addleson and Rice 1992).

angiospermous A major group of plants with closed seeds. This group includes all hardwoods (i.e. broad leafed trees). Compare with gymnopspermous* (Oliver *et al.* 1997).

anomaly An indication or manifestation (i.e. symptom) of a possible defect (CIB 1993).

arenaceous Sedimentary rocks in which the principal constituents are sand grains, including the various sorts of sands and sandstones (Oliver *et al.* 1997).

ascomycetes A major class of fungi consisting of about 30,000 described species (Solomon *et al.* 1996). Also called *sac fungi* because their sexual spores are produced in little sacs called *asci*. They include yeasts, powdery mildews and moulds. Compare with basidiomycetes* (Oliver *et al.* 1997).

basiodiomycetes A major class of fungi with about 16,000 species worldwide in distribution. Also called *club fungi*, because of their club-shaped basidium (i.e. the spore producing organ of basidiomycetes). Wood-rotting fungi are basidiomycetes (Singh 1993). Compare with ascomycetes* (Oliver *et al.* 1997).

bore dust (frass) The dust deposited by certain wood-boring insect larvae in their tunnels. It usually consists of chewed wood fragments and faecal pellets. These pellets comprise wood residues that have passed through the gut of the larvae. They have a shape and size that is often characteristic of the larval species concerned (Oliver *et al.* 1997).

brown rot Decay by wood-destroying fungi in which the cellulose component of the wood cell walls is broken down. The wood develops cuboidal cracking and is friable when handled (Oliver *et al.* 1997).

bossing The detachment but not dislodgment of fragments of plaster or render from a substrate because of loss of bond. (Contrast with spalling qv.)

building diagnostics The branch of Building Pathology that deals with the identification, diagnosis and prognosis of defects. It can also be used to describe investigations and analyses of problems with systems and equipment in a building.

building pathology   The systematic investigation and treatment of building defects, their causes, their consequences and their remedies (CIB 1993). It involves a holistic approach to understand how the various mechanisms by which the material and environmental conditions within a building can be affected (Watt 1999).

calcite   A crystalline form of calcium carbonate, often associated with igneous or sedimentary rocks (Oliver *et al.* 1997).

cambium   The zone in the trunk of a tree between the bark and the wood where secondary thickening and growth in girth of the tree take place (Oliver *et al.* 1997).

capillarity   Absorption of a liquid due to surface tension forces: for example, "rising damp" in walls is caused by capillary rise of the water in small pores of the walling material (CIB 1993).

capillary   A thin hollow tube. The extent to which water will rise in a capillary tube against the force of gravity is inversely proportional to the diameter of the capillary (Oliver *et al.* 1997).

capillary moisture (in a wall)   The moisture that rises in a wall as a result of rising capillarity in the fine cracks in the masonry, and especially in the mortar (Oliver *et al.* 1997).

carbonation   The transformation in concrete or cement mortar of the free alkali and alkali-earth hydroxides in the cement matrix into carbonates, due to attack by carbon dioxide in the atmosphere (CIB 1993).

cellulose   A polymeric constituent of plant cell walls composed of glucose units, often forming microfibril structures orientated as spirals, generally in the direction of the length of the cell and the grain of the wood. They confer much of the tensile strength in the cell wall and thus in the wood (Oliver *et al.* 1997).

chipboard   A wood-based sheet material in which wood particles are bonded together using a resin-based glue in a heated press (Oliver *et al.* 1997).

chlorophyll   The green pigment in plant leaves, which is essential for the photosynthetic process (Oliver *et al.* 1997).

clue   A piece of evidence that leads one towards the solution of a problem. Contrast with cue*.

combined water   The water that is present in wood cell walls. Once the moisture content drops below fibre saturation point, the combined water dries and some of it evaporates. This loss of water causes movement and reduction in electrical conductivity, but strength increases. If the moisture content increases again, the effects on these wood properties are reversed. Generally, it is the prevailing ambient humidity that determines the amount of combined water present. Thus, wood moisture content changes if humidity conditions alter (Oliver *et al.* 1997).

condensation   The precipitation of liquid from its vapour resulting from lowering of the temperature (at or below dew point) at a constant pressure; especially the disposition of water from warm moist air on to a relatively cold surface (CIB 1993).

condition   Level of critical properties of a building or its parts, determining its ability to perform (ISO 15686-1:2000).

condition survey   Inspection of a building or other structure, at a certain date, to determine its state of repair and any requirements for maintenance (Noy and Douglas 2005). It is often carried out on a regular basis for certain types of property, for example quinquennial inspection of church buildings (Dallas 2003). Usually it is presented in the form of a schedule of condition.

conservation   Preserving a building purposefully by accommodating a degree of beneficial change. It includes any "action to secure the survival or preservation of buildings, cultural artefacts, natural resources, energy or other thing of acknowledged value for the future" (BS 7913: 1998).

consolidation   Basic adaptation and maintenance works to ensure a building's ongoing beneficial use (Douglas 2006).

corrosion   The deterioration of a metal by chemical or electrochemical reaction with its environment. Occasionally used, incorrectly, to apply to non-metallic materials, for example concrete (CIB 1993).

crack   A linear discontinuity produced by fracture. Elongated narrow opening. Synonyms include break, split, fracture, fissure, separation, cleavage, etc. in various applications (CIB 1993).

crazing   A network of surface cracks. Used generally to describe surface cracking on concrete surfaces and paint film. Also specifically to describe the fine network cracking of ceramic glazes by, for example, differential thermal expansion between glaze and tile body, or moisture expansion of the body (CIB 1993).

creep   The slow deformation of a stressed material at temperatures which may be within or above the normal working range of the material (CIB 1993).

cryptoclimate   The climate around a building and its elements (Chandler 1989). Compare with macroclimate (the climate of a whole country of region); mesoclimate (the climate of a district or part of a country); and microclimate (the climate of a site).

cryptoefflorescence   The formation of soluble salts below the surface of a material. (e.g. limestone). This can cause exfoliation of the brick or stone (Addleson and Rice 1992).

crystal   A substance may become crystalline on drying or cooling, forming a structure in which all of the constituents are arranged in a regular three-dimensional pattern (Oliver *et al.* 1997).

cue   An intentional event used to initiate action. More appropriate to medical than building diagnostics.

damage   The harm, injury or impairment of value resulting from a failure. "Damages" is a legal term for the sum of money claimed or awarded in compensation for loss or injury (Houghton-Evans 2005).

dampness   Condition of being slightly wet – usually not so wet that liquid water is evident–such as wetness caused by condensation on a porous substrate or water transmitted up a porous wall by capillary action (CIB 1993). Dampness can be said to occur when an atmosphere or material is wetter than 85 per cent RH (Oliver *et al.* 1997).

desorption   This is a form of outgassing of moisture and is the reverse process to adsorption (Oliver *et al.* 1997).

decision tree   A decision tree is a graphical tool for helping a decision-maker to choose between several courses of action, and indicates their possible consequences (including resource costs and risks). It's constructed to help with making decisions to create a plan to reach a goal (see examples in Chapter 2).

decomposition   The separation of a material into elements or parts (CIB 1993).

dehydration   The removal of water by heating, chemical action or other means (Oliver *et al.* 1997).

degradation   Changes over time in the composition, microstructure and properties of a component or material, which reduce its performance (ISO 15686-1:2000). It also usually refers to the deleterious effects of the weather such as sunlight, in particular on organic materials. Strictly speaking, degradation is defined as the conversion of a complex molecule into simpler fragments.

degradation agent   Whatever acts on a building or its parts to adversely affect its performance (ISO 15686-1:2000).

degradation mechanism   Chemical, mechanical or physical path of reaction that leads to adverse changes in a critical property of a building product (ISO 15686-1:2000).

delamination   The breakdown of a material by separation of the layers of which it is composed (CIB 1993).

density   Expressed as the mass or weight of a unit volume of a substance in kilograms per cubic metre (kg/m^3). Porous substances can have a wide range of densities depending on the volume of pores, e.g. wood and aerated concrete products. Generally, the highest densities are associated with the greatest strength (lead being an exception).

design life   The period of time over which a building or a building subsystem or component (e.g. roof, window, plumbing) is designed to provide at least an acceptable minimum level of performance (Iselin and Lemer 1993). It is the period of use as intended by the designer – for example, as stated by designer to the client to support specification decisions (BS 7543).

desiccation   The drying out of a material such as clay soil.

deterioration A reduction in ability to perform up to the anticipated standard (CIB 1993).

dew point The temperature at which condensation occurs.

diagnosis An impartial assessment of all the data and evidence available to determine the root cause of a problem. Deciding the nature of a fault from its symptoms (Watt 1999). An investigation or analysis of the cause or nature of a condition, situation or problem (Webster's New Collegiate Dictionary).

diagnostic tree This is a procedural guide for the diagnostic enquiry, which unlike a fault tree gives an answer on the actual cause(s) of a certain event (CIB 1993). It is used for a specific situation where the actual branches of the fault tree have already been determined. (See example of diagnostic tree in CIB 1993.)

diffusion The general transport of matter whereby molecules or ions mix through normal thermal agitation. It is the movement of substances from a centre of high concentration to achieve a balanced concentration (equilibrium). This process is an example of the high-low principle (Addleson and Rice 1992).

dilapidations According to Hollis (1988), "dilapidations" refers to a state of disrepair in a property where there is a legal liability for the condition of disrepair. It is normally presented in the form of a schedule of dilapidations.

dry rot Decay in buildings by the fungus *Serpula lacrymans*. The name in one sense is apt since the fungus can cause decay of wood at relatively low moisture content (down to 20%), although this is still wet in terms of normal equilibrium moisture contents of timbers in buildings.

drying The process of removal of water from a material. It can be achieved by air drying, kiln drying or by the use of a dehumidifier.

durability The ability of a building or its parts to perform its required functions over a period of time and under the influence of internal and external agencies or mechanisms of deterioration and decay (Watt 1999). It is also a measure of a building's ability to resist deterioration. "Durability is not an inherent property of a material or component, although the term is sometimes erroneously used as such" (ISO 15686-1:2000).

economic life The period of time over which costs are incurred and benefits or dis-benefits are delivered to an owner; an assumed value sometimes established by tax regulations or other legal requirements or accounting standards not necessarily related to the likely service life of a facility or subsystem (Iselin and Lemer 1993).

efflorescence Soluble salts brought to the surface of building materials (such as brick and concrete) by drying (CIB 1993).

electro-osmosis The movement of a liquid under an applied electric field through a fine tube or membrane (Oliver *et al.* 1997).

emulsion The suspension of a substance in a liquid in which it cannot dissolve because it is immiscible* (Oliver *et al.* 1997).

enzymes Complex protein chemicals produced by living organisms that cause biochemical reactions to take place. They are catalysts produced by living cells. Wood-destroying fungi exude enzymes to digest wood cell walls as part of the decaying process (Oliver *et al.* 1997).

equilibrium moisture content Many porous building materials have a moisture content which is normally in balance with the prevailing moisture conditions. The moisture content will be greater with materials of higher hygroscopicity (Oliver *et al.* 1997).

ettringite A crystalline inorganic mineral formed in cement during the hydration of calcium aluminates in the presence of sulphate ions (Oliver *et al.* 1997).

eukaryote Organism whose cells possess nuclei and other membrane-bounded organelles. Compare with prokaryote (Solomon *et al.* 1996).

evaporation The conversion of a liquid into vapour, at a temperature below the boiling point. The rate of evaporation increases with rise of temperature, since it depends on the saturated vapour pressure of the liquid, which rises until it is equal to the atmospheric pressure at the boiling point (Oliver *et al.* 1997).

evidence The available body of facts, data and information that has factual significance – essential for establishing the proof of something.

exfiltration Leakage out of a material or structure (CIB 1993).

failure The loss of the ability of a building or its parts to perform a specified function (ISO 15686-1:2000). It can also be classed as the consequence or effect of a defect or fault.

fatigue The weakening of a material caused by repeated or alternating loads (CIB 1993).

fault State characterised by an inability to perform a required function, excluding the inability during preventative maintenance or other planned actions, or due to a lack of external sources (Watt 1999). An unexpected deviation from requirements, which would require considered action regarding the degree of acceptability. It is also considered as a departure from good practice or contravention of authoritative standards.

fault tree According to CIB (1993) a fault tree can be described as a cognitive model or diagram which represents an outline of the possible errors affecting the performance of a building system or component. The tree can indicate correlated phenomena and chain of events leading from error to failure. (See examples of fault tree in CIB 1993).

fibre A thin wood cell up to 3 mm long with pointed ends, orientated along the grain and having a mechanical strength function (Oliver *et al.* 1997).

fibre saturation point A moisture content of wood in the range 25 to 30%, when (theoretically) all the free water in the cell lumen has been lost as a result of drying but the combined water in the cell walls still remains (Oliver *et al.* 1997).

fungus A form of plant growth obtaining its nutrition by breakdowns of organic matter, usually associated with the presence of dampness, e.g. in timber. The plants are characterised by the absence of chlorophyll* (CIB 1993). Technically classified as a heterotrophic* eukaryote* with chitinous cell walls and a body usually in the form of a mycelium* of branched, thread-like hyphae*, or unicellular (Solomon *et al.* 1996). Most fungi are decomposers (i.e. saprophytes); some are parasitic. Fungi are nucleated organisms which lack green colouring matter or chlorophyll*, and therefore cannot photosynthesise their food. Basidiomycetes* are the subgroup of fungi that are responsible for wood-rooting species causing dry rot and wet rot.

grain The direction of the axes of the principal cells in wood. Greatest tensile strength and permeability are to be found in the grain direction. Movement is greatest at right angles to the grain. The end of a piece of converted timber is called the end grain, and the side is the side grain. Permeability to liquids is greatest along the grain, penetrating from exposed end-grain areas (Oliver *et al.* 1997).

green timber A log, a piece of lumber or freshly converted timber which has a high moisture content (50–200%) because it has not been dried (Oliver *et al.* 1997).

growth rings Caused by alterations or the cessation in the growth of the cambium; often associated with seasonal growth patterns. Different forms of wood cells are produced and the variations may be visible as a series of rings on the end grain (Oliver *et al.* 1997).

gymnospermous A major group of plants with open seeds. This group includes all softwoods (i.e. needle-leafed trees). Compare with angiospermous* (Oliver *et al.* 1997).

hardwood Timber from a broadleaf tree (an angiospermous* plant) (Oliver *et al.* 1997).

heartwood The central zone of the trunk of a tree where active sap conduction and transpiration has ceased. On converted wood, heartwood is often darker coloured, and sometimes more resistant to decay and insect attack and less permeable to liquids and wood preservatives (Oliver *et al.* 1997).

heave Upward movement of soil (ground) or of a structure which it supports (CIB 1993). May be caused by rehydration* of clay soil or sulphate attack in hardcore.

heterotroph Organism that cannot synthesise its own food from organic raw materials and therefore must live either at the expense of other organisms or upon decaying matter (i.e. includes most bacteria and fungi). Compare with autotroph (i.e. organism that can synthesise organic compounds from raw materials – includes some bacteria).

Heterotroph is also called consumer; autotroph is called producer (Solomon *et al.* 1996).

honeycombing Usually used to refer to a defective condition in concrete. The concrete contains interconnecting large voids due to loss or lack of mortar (CIB 1993).

hydrolysis Any irreversible chemical reaction initiated by or involving water (Oliver *et al.* 1997).

hydrophobic A substance or surface that is repellent of water. Compare with hydrophilic* (Oliver *et al.* 1997).

hydrophilic A substance or surface that is attracted to water (Oliver *et al.* 1997).

hygrometer An instrument for measuring atmospheric humidity (Noy and Douglas 2005). There are basically three types of hygrometer in use: the whirling hygrometer; the hair hygrometer; and the digital electronic hygrometer.

hygroscopic If a substance is hygroscopic, it has an attraction to water. Hygroscopic substances may absorb or lose water to the environment as they attempt to achieve a balance with the prevailing atmospheric relative humidity (Oliver *et al.* 1997).

hygroscopic moisture The moisture in a material that can be attributed to the hygroscopic materials present. Thus for wood, the cellulose attracts atmospheric moisture; in masonry, hygroscopic salts act in a similar fashion. The amount of hygroscopic moisture will be related to the prevailing ambient relative humidity in the surrounding atmosphere (Oliver *et al.* 1997).

hyphae Tubular branching root-like structures that are produced by a fungus. They are microscopic in size, and penetrate into the substance on which the fungus is growing. The hyphae of wood-destroying fungi produce the enzymes which digest and decay the wood substance (Oliver *et al.* 1997).

igneous rock A rock deriving directly from volcanic activity (Oliver *et al.* 1997).

immiscible Incapable of being mixed (Oliver *et al.* 1997).

insoluble Incapable of being dissolved (Oliver *et al.* 1997).

insulation Any means of confining a transmission phenomenon, e.g. of heat or sound, to obviate or minimise loss or damage (Oliver *et al.* 1997).

infiltration Leakage into a material or structure (CIB 1993).

integrity In Building Pathology: soundness, with no part or element deficient in performance (CIB 1993).

interstitial Occurring within the thickness of some material element – usually used in the context of "interstitial condensation", which means condensation that occurs within the thickness of a building element or within its component materials (CIB 1993).

kiln drying Drying of a material by artificial means under high heating conditions. Timber can be dried in a kiln in which, by controlling air flow, humidity and temperature, it is possible to dry wood to the low moisture levels necessary in centrally heated buildings. Such kiln-dried wood, when dried to the appropriate equilibrium moisture content, should not exhibit movement in use – unless, of course, moisture conditions alter again (Oliver *et al.* 1997).

lignin A phenolic-type polymer substance present in wood cells that binds the fibres together (Oliver *et al.* 1997).

life cycle The sequence of events in planning, design, construction, use, and disposal (e.g. through sale, demolition, or substantial renovation) during the service life of a facility; may include changes in use and reconstruction (Iselin and Lemer 1993).

life-cycle cost The present value of all anticipated costs to be incurred during a facility's economic life; the sum total of direct, indirect, recurring, nonrecurring, and other related costs incurred or estimated to be incurred in the design, development, production, operation, maintenance, support, and final disposition of a major system over its anticipated life-span (Iselin and Lemer 1993).

loss The consequences of a defect or failure, expressed in terms of costs, injuries, loss of life, etc. (CIB 1993).

macropores Fine or small cavities in a material – between groups of particles (Oliver *et al.* 1997).

maintenance A "combination of all technical and administrative actions, including supervision actions, intended to retain an item in, or restore it to, a state in which it can perform a required function" (BS 3811: 1993). Maintenance involves routine work necessary to keep the fabric of a building, the moving parts of machinery etc., in good order (BS 7913: 1992). The word comes from the French verb "maintenir", which means to hold (Chudley 1983).

metamorphic rock A rock formed as a result of the action of heat and pressure on sedimentary or igneous rock (Oliver *et al.* 1997).

micropores Minute cavities in a material – between particles (Oliver *et al.* 1997).

moisture content The amount of water present in a substance expressed as a percentage of its oven dry weight. In timber, it can be measured reasonably accurately with an electrical moisture meter (Oliver *et al.* 1997).

moisture movement Many building materials may show movement, especially shrinkage on initial drying out of moisture. If the material has a high degree of hygroscopicity, swelling movement may occur if the material increases in moisture content (Oliver *et al.* 1997).

mould A superficial growth of fungus that occurs under conditions of high humidity. The minimum level for mould growth is frequently quoted as 70 per cent RH (Garrat and Nowak 1991).

mycelium A mass of visible hyphae produced by a fungus which can be seen on the surface of timber or masonry (Oliver *et al.* 1997).

mycology The study of fungi (from the Greek *mykes*, mushroom or fungus; *logos*, discourse) (Oliver *et al.* 1997).

myxomycota The group of wall-less fungi (e.g. slime moulds). Compare with eumycota (the true walled fungi, such as basidiomycetes*) (Oliver *et al.* 1997).

natural durability The inherent resistance of certain timbers to fungal decay. The classification of natural durability is based on the results of ground contact graveyard type tests (Oliver *et al.* 1997).

obsolescence The condition of being antiquated, old fashioned, or out of date, resulting when there is a change in the requirements or expectations regarding the shelter, comfort, profitability, or other dimension of performance that a building or building subsystem is expected to provide. Obsolescence may occur because of functional, economic, technical, or social and cultural change (Douglas 2006).

osmosis This is diffusion of a solvent through a semi-permeable membrane into a more concentrated solution, tending to equalise the concentrations on both sides of the membrane (Oliver *et al.* 1997).

pargetting Coating of cement mortar in flues or between the underside joints of slates or tiles (Oliver *et al.* 1997).

performance The degree to which a building or other facility serves its users and fulfils the purpose for which it was built or acquired; the ability of a facility to provide the shelter and service for which it is intended (Douglas 2006). It is a quantitative expression of behaviour of an asset or product in use (BS 6019).

permeability The rate of diffusion of a gas or a liquid under a pressure gradient through a porous material. The permeability of wood describes the ease or difficulty of the flow of liquids through the wood. Greatest permeability is found along the grain direction (Oliver *et al.* 1997).

photosynthesis The process carried out in green leaves of plants whereby carbon dioxide and water are combined in the presence of sunlight to form glucose, which is the basis of all the other products needed by the plant (Oliver *et al.* 1997).

physical service life The time it takes for a building, subsystem, or component to wear out. It is the time period after which a facility can no longer perform its function because increasing physical deterioration has rendered it useless (Douglas 2006).

plaster Internal gypsum-based coating of walls and ceilings (Oliver *et al.* 1997).

plasticiser An additive or admixture to cementitious materials which improves workability (Oliver *et al.* 1997). Plasticisers able a reduction in the water/cement ratio and can provide higher strength, lower shrinkage and greater frost resistance.

plywood   A composite sheet material composed of wood veneers bonded with adhesive. Marine quality plywood contains water-resistant glues. Usually, the board is made of veneers in which the grain is laid at right angles to that of the adjacent ply to provide redistributed strength properties and restriction of movement (Oliver *et al.* 1997).

pointing   The process of raking out old mortar from masonry, and refilling with a compatible mortar (Oliver *et al.* 1997).

polymerisation   The combination of several broadly similar molecules to form a more complex molecule with the same basic formula as the simple molecule (Oliver *et al.* 1997).

porosity   The percentage of space in a material (Oliver *et al.* 1997).

preservation   Arresting or retarding the deterioration of a building or monument by using sensitive and sympathetic repair techniques. Preservation means "the state of survival of a building or artefact, whether by historical accident or through a combination of protection and active conservation" (BS 7913: 1998). It also can be defined as "the act or process of applying measures necessary to sustain the existing form, integrity, and materials of an historic property (Weeks and Grimmer 1995). Preservation focuses on the maintenance and repair of existing historic materials and retention of a property's form as it has evolved over time. It includes protection and stabilisation measures (Douglas 2006).

prognosis   An impartial, technical assessment of the probable projection and outcome of a building deficiency. Predicting or forecasting the course of a fault from its symptoms (Watt 1999).

protection   The legal use of this term involves the provision of legal restraints or controls on the destruction or damaging of buildings and so on, with a view to ensuring their survival or preservation for the future (Douglas 2006). Physical protection may be either temporary (e.g. tarpaulins over an exposed roof surface undergoing refurbishment) or permanent (e.g. over-roofing scheme).

redundancy   The consequence for a building when it becomes superfluous or excess to requirements (Douglas 2006). It is often triggered by obsolescence.

rehydration   The absorption of moisture in a material owing to excessive wetting.

relative humidity   This is the existing water vapour pressure of the atmosphere expressed as a percentage of the saturated water vapour pressure at the same temperature (Oliver *et al.* 1997). RH is similar to the saturation percentage, but is a better measure of moisture in the atmosphere because it indicates how near the quantity of water vapour present in the air is to saturation under the same conditions. The basic formula for RH is:

$$RH = \frac{\text{Actual water content}}{\text{Saturated water content}} \times \frac{100}{1}$$

render   A cementitious coating applied to the external surface of masonry to improve appearance and weather resistance (Oliver *et al.* 1997).

retro-fit   A modern engineering term imported from North America. To retro-fit is to introduce a modern component into a redundant or out-moded subject to enhance, revitalise or indeed, reinstate its original effectiveness (Oliver *et al.* 1997).

rising dampness   Moisture rising up masonry walling or solid flooring as a result of capillary action (see Massari and Massari 1993). True rising dampness is relatively rare – when it does occur, the construction usually lacks a damp proof course/membrane. Bridging moisture is the main cause of indirect rising dampness.

salt   A substance derived from the action either between an acid and an alkali or an acid and a metal. Salts are generally crystalline when dry, but form ions when dissolved in water (Oliver *et al.* 1997).

sapwood   The outer zone of wood in the tree where active conduction of sap occurs. In converted timber, the sapwood is usually light coloured, less resistant to decay, more susceptible to insect attack and, when dry, more permeable to wood preservatives (Oliver *et al.* 1997).

sarking   An internal lining in a pitched roof under tiles or slates (Oliver *et al.* 1997).

seismic staining   Uncontrolled flow of rainwater under window sills, string courses, parapets, edge beams, and so on, giving an irregular seismic pattern of staining on wall surfaces (Parnham 1996). Lack of adequate drips or discontinuity of drip to underside of sills and so on is one of the main causes of this problem.

service (or working) life   Actual period of time during which no excessive expenditure is required on operation, maintenance or repair of a component or construction – as recorded in use (BS 7543).

settlement   A downward movement of a building, caused by above-ground factors (contrast with "subsidence" qv).

sick building syndrome   sick building syndrome (SBS) is a phenomenon in which some or many occupants of a property complain of ailments directly resulting from using the building (Douglas 2006). Not everybody who works in such a building will necessarily become sick. The main symptoms associated with SBS are dryness of skin, eyes, nose and throat; allergic effects – watery eyes, runny nose; asthmatic effects – chest tightness, wheezing; and, general ailments such as lethargy and headache.

sign   Any objective evidence of a defect.

spalling   The detachment of fragments, usually of flaky shape, from a larger mass by a blow, or by the action of weather or pressure (CIB 1993).

specific humidity   This is a measure of the mass of water vapour per unit mass of moist air. It is also known as the "mixing ratio", and is expressed in kg/kg of dry air involved (Oliver *et al.* 1997).

softwood Timber from coniferous cone-bearing trees, which are usually evergreen and have needle-shaped leaves (a gymnospermous plant*) (Oliver *et al.* 1997).

soluble A substance that dissolves in a liquid is soluble in that liquid (Oliver *et al.* 1997).

solvent A liquid in which other substances are dissolved. The movement of the solvent may cause the dissolved substances to be carried to other places where they may be deposited if the solvent evaporates. Solvents are used to carry the active ingredients of wood preservatives or damp proofing chemicals into wood or masonry. Penetrating damp can act as a solvent for mineral salts present in masonry (Oliver *et al.* 1997).

spore A reproductive fungal cell, microscopic in size, which is produced by the fungus and serves to distribute the species to a fresh substrate or to enable the fungus to survive unfavourable environmental conditions (Oliver *et al.* 1997).

sporophore A spore-producing structure (i.e. a fruiting body) of a fungus (Oliver *et al.* 1997).

strands Conducting structures produced by the mycelium of certain wood-destroying fungi that assist in the conduction of water and other essential substances to the growing fringe of the mycelium. Sometimes referred to as rhizomorphs (Oliver *et al.* 1997).

stabilisation Substantial adaptation works to ensure a building's long-term beneficial and safe use. It often includes strengthening works such as stitching and underpinning (Douglas 2006).

stain To discolour. A discolouration (CIB 1993).

subsidence A downward movement of a building, caused by below-ground factors (contrast with "settlement").

subsystem Functional part of a system, and often used interchangeably with that term – for example, heating subsystem being part of HVAC system (Iselin and Lemer 1993).

summer (or reverse) condensation This phenomenon can occur in summertime within certain wall constructions (BRE IP 12/88). In solid walls with insulated linings orientated from east-south-east through south to west-south-west, sunshine falling on damp masonry can drive water vapour into the construction, causing condensation on the outside face of the vapour control layer (Oliver *et al.* 1997). Because the vapour control layer is always supposed to be placed on the warm side of construction, any insulation behind it would in these circumstances become wet. This form of condensation, however, has apparently not been a problem in insulated "no-fines" construction (Oliver *et al.* 1997).

surface tension Property of liquid surfaces to assume minimum area and, in so doing, liquid surfaces exhibit certain features resembling the properties of a stretched elastic membrane (Addleson and Rice 1992).

sustainability A set of processes aimed at delivering efficient built assets in the long term (Douglas 2006).

symptom Any subjective evidence of a defect.

system Collection of subsystems, components, or elements that work together to provide some major aspect of shelter or service in a constructed facility (e.g. plumbing system, heating system, electrical system, and roofing system). Also, a set of building components specifically designed to work together to facilitate construction, such as an integrated building system (Iselin and Lemer 1993).

tanking The application of an impervious material to a wall to prevent penetrating, usually in below-ground situations (Oliver *et al.* 1997).

tracheid A fibre-shaped wood cell up to 3 mm long, orientated along the grain, that performs both mechanical and conductive functions. It constitutes the major wood tissue in softwoods (Oliver *et al.* 1997).

technology This can be defined simply as the systematic study and application of how artefacts are made and used (Douglas 2006).

thermal pumping It's a mechanism that can cause water penetration to sheeted roofs. "A physical process in which air or water is drawn through a material bounding an air-filled space by the vacuum caused as a result of pressure changes due to changes in temperature. This can occur where the joints between the sheets are tightly lapped but not completely airtight, and where there is a cavity below the sheets which is sealed to air except at the laps. When the roof is subject to sudden changes in temperature, water which is held in the laps by capillarity can be sucked over them" (Addleson and Rice 1992).

thermal shock The force, arising out of thermal expansion or contraction, which causes disruption of a material on sudden heating or cooling (CIB 1993).

thermoplastic plastics Class of plastics which can be softened and re-softened indefinitely by the application of controlled heat and pressure (Addleson and Rice 1992).

thermosetting plastics Class of plastics which undergoes a chemical reaction during the hardening process and cannot subsequently be reshaped by the application of controlled heat and pressure (Addleson and Rice 1992).

turnerization The term used to describe a proprietary system of applying a bituminous coating to the external surface slated or tiled roofs to enhance their weathertightness (Douglas 2006).

vapour check A layer which checks water vapour movement and so prevents or minimises water condensing in an insulant around a cold surface (Oliver *et al.* 1997). It should always be placed on the warm side of construction.

void (in the context of cellular materials, such as concrete) A cavity unintentionally formed and substantially larger than the characteristic individual cells (CIB 1993).

water/cement ratio The strength of concrete and other cementitious products depends on the ratio of the weight of water to the weight of cement: the lower the ratio, the higher the strength of the concrete. The minimum ratio for a workable mix is 0.45, but 0.25 is sufficient to set the cement (Oliver *et al.* 1997).

water-repellency The ability of a substance to repel water. A liquid water repellent can be applied to walls to encourage drying by preventing further water uptake. Water repellents are usually hydrophobic (Oliver *et al.* 1997).

water vapour pressure That part of atmospheric pressure due to water vapour (Oliver *et al.* 1997).

weather To degrade under the action of the weather. Also used to describe the inclusion of a slight slope to throw off rainwater, for example on a sill (CIB 1993).

weathering Action of water in producing degradation which a material undergoes when it is stressed beyond its elastic limit (CIB 1993).

wet rot The decay of timber by one of several different species of wood-destroying fungi that require a relatively high timber moisture content (usually 30% minimum) (Ridout 1999). (Excludes dry-rot due to *Serpula lacrymans.**)

white rot A category of fungus that can decay both the lignin and the cellulose of the wood cell wall (Ridout 1999). Characteristically, the wood is bleached, turns an off-white, and becomes soft and springy on the inside even though the outer surface appears sound. Cuboidal cracking does not develop as extensively as with dry rot. All white rots are wet rots.

whole life cycle cost Generic term for the costs associated with owning and operating a facility from inception to demolition, including both initial capital costs and running costs (Watt 1999).

woodworm A popular term that describes all forms of insect damage to timber (Oliver *et al.* 1997).

yield The permanent deformation that a material undergoes when it is stressed beyond its elastic limit (CIB 1993).

Schedule of defects

Item No.	Element	Defect	Cause	Effect	Cost or Priority Rating
1.0	ROOF				
1.1	Coverings	Loose and broken slates	Nail sickness	Rainwater penetration Fungal attack	£12,000 (2)
	Skylights	Missing flashing	Storm damage	Ditto	2 × £1,000 (1)
	Cappings	Ridge capping missing	Storm damage	Ditto	
2.0	WALLS				
2.1	Rendering				

Notes:
Priority ratings:

1. Currently critical – attend within 1 week
2. Potentially critical – attend within 1 month
3. Desirable – attend within 6 months
4. Recommended – attend within 1 year

Appendix C

Various defects data analysis checklists

General Dampness Assessment Checklist

DAMPNESS ASSESSMENT			
PROPERTY:			
SURVEYOR:........................DATE:..................................			
	Yes	No	Remarks
Affected Elements Floor: Walls: Ceiling: Other:.................			
Building Conditions Solid masonry: Cavity walls: Pitched and/or flat roof: Well maintained: Empty:			
Symptoms of Attack Tide marks: Efflorescent salts: Mould/algae: Sweating surfaces: Ponding: Other:.................			

(Continued)

	Yes	No	Unsure	Remarks
Moisture Levels ERH of air: WME of affected wall: WME of timbers: Speedy meter used: Other:..................				
Dampness Factors DPC present: DPC defective: High outside ground level Other:..................				
Likely Source of Moisture Leaking external pipe: Leaking internal pipe: Rainwater: Condensation: Bridging/rising damp Other:..................				
Site Notes:				

Fabric Dampness Data Checklist

FABRIC DAMPNESS DATA

PROPERTY:...

LEVEL: .... ROOM:.............................

SURVEYOR:................................... DATE:.............................

Wall I = Int. E = Ext.	Depth of sample in mm and height above structural floor level								
	0–5 mm			5–50 mm			> 50 mm		
	0–0.1 m	0.1–0.5 m	0.5–1 m	0–0.1 m	0.1–0.5 m	0.5–1 m	0–0.1 m	0.1–0.5 m	0.5–1 m
1a (I)									
1a (E)									
1b (I)									
1b (E)									
1c (I)									
1c (E)									
1d (I)									
1d (E)									
Site Notes:									

Moisture Content of Masonry Samples Checklist

				MOISTURE TEST RESULTS		
PROPERTY:						
SURVEYOR:........................ DATE:..............................						
Sample No.	Location	Wall	Approx. Height & Depth	*Moisture Contents %		
				"Speedy"	Oven	"Protim."
1						
2						
3						
4						
5						
6						
7						

Site Notes:
- At least two of the three methods listed at * above should be used to determine the moisture level in a sample.

Condensation Assessment Checklist

				CONDENSATION DATA			
PROPERTY:...							
SURVEYOR:............................. DATE:...							
Item No.	Location	Outdoor Air Temp. C	Indoor Air Temp. C	Surface Temp. C	Ave. ERH %	Dew Point Temp. C	Remarks
1							
2							
3							
4							
5							

Site Notes:
- Dew-point temperature is established using either a whirling (hand-held analogue-type) hygrometer and psychrometric chart or by using an electronic hygrometer to give a more direct result.
- Remarks column should indicate condensation risk as high/medium/low.

Fungal & Insect Attack Checklist

FUNGAL AND INSECT ATTACK			
PROPERTY:...			
SURVEYOR:................................. DATE:.............			
	Yes	No	Remarks
Locus of Attack			
Basement:			
Ground Floor:			
Middle-Floor:			
Top Floor:			
Roof Space:			
Affected Elements			
Floor:			
Walls:			
Ceiling:			
Other:.................			
Symptoms of Attack			
Fruiting Bodies:			
Spore dust:			
Mycelium:			
Boreholes:			
Bore frass:			
Slime/mould:			
Decayed wood:			
Condition of Cracking			
Sharp edges:			
Ragged edges:			
Clean:			
Dirty:			

	Yes	No	Unsure	Remarks
Building Conditions				
Solid masonry:				
Cavity walls:				
Pitched &/or Flat roof:				
Well maintained:				
Empty:				
Environmental Factors				
Temperature:				
Humidity:				
Ventilation:				
Light:				
Likely Type of Outbreak				
Dry Rot:				
Wet Rot:				
Non-woodrotting:				
Woodworm:				
Other:.................				
Site Notes:				

Crack Damage Checklist

CRACK DAMAGE			
PROPERTY:			
SURVEYOR: .... DATE:			

	Yes	No	Remarks
Locus of Problem			
Basement:			
Ground floor			
Middle floor			
Top floor			
Roof space			
Other:			
Style of cracking			
Horizontal:			
Vertical:			
Diagonal:			
Random:			
Other:			
Size of Cracking			
0.1–1 mm:			
1–2 mm:			
2–5 mm:			
5–15 mm:			
15–25 mm:			
> 25 mm:			
Condition of Cracking			
Sharp edges:			
Ragged edges:			
Clean:			
Dirty:			
Other:			

	Yes	Possible	No	Notes
Suggested Cause/s of Cracking				
Alterations:				
Settlement:				
Subsidence:				
Sulphate attack:				
Wall-tie failure:				
Combination:				
Other:				

Significance of cracks (decision matrix)	Is the movement across the crack:			
	Static?	Cyclic?	Prog?	
Is the crack only aesthetic?				
Is the crack affecting serviceability?				
Is the crack affecting stability?				

Sample diagnostic report

TECHNICAL REPORT

Reference	X005/JB/05
Date	30 September 2005
From	J. Bloggs, Building Diagnostics Co Anytown Britain
To	Ms T. Client 20 Acme Crescent Anytown Britain
Subject	**ASSESSMENT OF DAMP KITCHEN FLOOR IN DWELLING, AT 20 ACME CRESCENT, ANYTOWN, BRITAIN**

1.0 Introduction

1.1 This is a report on the damp floor screed in the kitchen of the dwelling owned by Ms Client at 20 Acme Crescent, Anytown, Britain. The aim is to help establish the source and significance of the problem.

1.2 The investigation and report was undertaken by Mr J. Bloggs of Building Diagnostics Co. in Anytown, on the instructions of Ms Client. A visit to the property to inspect the problem was made by Mr Bloggs on the morning of Thursday 1 September 2005. On the day of the inspection the weather was fair and mild.

1.3 The writer used a "Protimeter *Surveymaster SM*" electronic moisture-reading meter to determine the presence of excess moisture in the floor screed. The instrument has two modes of operation: a standard two-pin "measure" mode for surface dampness and a non-destructive radio frequency "search" for below-surface dampness. The latter mode was used in this investigation.

1.4   Although calibrated for timber, the *Surveymaster SM* can give a meaningful indication as to whether or not porous non-wood materials are dry or damp. It has three main moisture zones: green (safe); amber (borderline); and red (danger).

1.5   This report shall be for the private and confidential use of the client for whom it is prepared and should not be relied upon by third parties for any use without the express written authority of the writer.

2.0   Investigation and analysis

2.1   Background

2.1.1   The property in question is a detached two-storey high quality dwelling which was completed in early May 1998. The client moved into the house on 29 May.

2.1.2   The ground floor is of suspended precast concrete construction. It consists of precast concrete tee beams with lightweight concrete infill blocks. The floor surface is finished with a cement screed. As drawings were not available the writer is unable to confirm the precise construction of the floor finish. In any event, the screed would be typically at least 50 mm thick if unbonded, or 65 mm thick for a floating screed on insulation.

2.2   Examination

2.2.1   The vinyl floor sheeting used as a temporary covering was removed to allow full access to the floor screed's surface. Some slight sweating was evident on the back of the flooring. The surface of the screed felt damp and cold to the touch.

2.2.2   The writer took several "Protimeter" readings using the radio frequency facility on the instrument at selective positions over the floor screed in both the kitchen and adjoining utility room. In every case the instrument showed the readings in the red zone. This indicated that the screed was still wet below the surface.

2.3   Discussion

2.3.1   Construction moisture can account for dampness problems in the initial occupancy of a new building. In this case the floor screed would only have had less than about 6 weeks to dry out. This coupled with the relatively wet weather conditions would explain the high residual moisture levels in the screed.

2.3.2   A wet screed would account for the sweating noted on the underside of the vinyl flooring and the mould growth on the back of the skirting boards. The plasterwork immediately above the skirting showed "dry" readings. Mould growth does not occur in cases of ground-borne damp because this type of moisture is contaminated with soil salts.

2.3.3   According to BS 5325 (1996): "For cement-sand laid directly over a damp proof membrane, one day should be allowed for each millimetre of thickness for the first 50 mm, followed by an increasing time for each millimetre above this thickness. It is thus reasonable to expect a screed 50 mm thick drying under good conditions, to be sufficiently dry in two months."

2.3.4   Ideally, cement screeds and concrete slabs should have a moisture content less than 75% before any flooring is laid onto them (BS 8203: 2001). This can be best ascertained by using the hygrometer method (see Appendix A).

2.3.5   The following extract from a recent Building Research Establishment (BRE) report (No. BR 332) on floors and flooring (1997) neatly describes the problems associated with precast concrete ground floors:

(Precast) concrete ground floors are normally constructed when the walls are at DPC level before the building is made watertight. It is possible, therefore, for rain to saturate the blocks or other components of the floor. Subsequently this leads to a long drying time before moisture sensitive flooring can be installed; it can often hold up completion and occupation, or lead to premature floor failure. It is therefore good practice, in the BRE's view, that a DPM or vapour control layer should always be placed above the structural floor and below the screed or timber panel product to control the upwards transmission of construction moisture. It is also vital to provide a turn-up to the DPM, otherwise moisture from accumulated rainwater can migrate up the inner leaf of the external wall.

3.0   Conclusions and recommendations

3.1   Conclusions

3.1.1   There was nothing to suggest that penetrating or rising dampness was the cause of the wet floor. The general coverage of the dampness in the floor screed also ruled out a leaking service pipe as a possible cause. It is possible, however, that moisture condensing on the unlagged water supply pipe has contributed to the more distinct damp patch at the edge of the screed below the sink.

3.1.2   The short time between the laying of the floor screed and occupancy of the house and the overall spread of dampness in the screed strongly suggests that construction moisture is the source of the problem in this case. Moreover, the premature laying of the vinyl floor covering would have inhibited the natural drying out of the screed.

3.1.3   The mould growth behind the skirtings is a result of condensation forming on the damp (cold) screed. This is a superficial problem which is unlikely to cause any long-term damage to the skirtings or surrounding construction. However, persistent mould growth does pose a health risk as it increases the release of toxic mould spores into the room.

3.1.4   It is understood that the client is intending to lay a high quality thin sheet wood flooring on an underlay in the kitchen. There is a risk that any floor-covering will fail, however, if the screed has not been allowed to dry out properly.

3.2   Recommendations

3.2.1   Dehumidifiers should be used for at least a week to help dry out the screed in the kitchen and utility room. During this process the vinyl sheet covering should be uplifted to allow the screed to breath and all windows should be kept shut to maximise the efficiency of the dehumidification. Additionally, the heating needs to be kept on as normal.

3.2.2   A two-coat surface DPM such as "Tremco *ES3000*" should be applied to the floor screed to suppress any residual moisture. This would allow the new flooring to be laid even if the relative humidity of the screed has not dropped below the 75% threshold by the end of the 1-week dehumidification period.

3.2.3   Before refixing them the back of the MDF skirtings should be cleaned using a fungicidal solution (such as Dulux's "Fungiwash") to remove the mould growth.

3.2.4   Ensure all bare copper water supply pipes are fully protected against condensation and frost action below the sink unit.

Joe Bloggs, BSc, MSc   Building Diagnostics
                                                  September 2006

References

Addleson, L (1992) *Building Failures: A guide to diagnosis, remedy and prevention* (3rd edition). London: Butterworth-Heinemann.

Addleson, L and Rice, C (1992) *Performance of Materials in Building.* Oxford: Butterworth-Heinemann.

Asbestos Working Party (2003) *Asbestos – RICS Guidance Note.* London: RICS Books.

ASHRAE (2005) "The commissioning process", *Guideline 0-2005*, American Society of Heating, Refrigeration and Air-Conditioning Engineers Inc., Atlanta GA.

Ashworth, A (1999) *Cost Studies of Buildings* (3rd edition). London: Longmans.

Atkins S and Murphy, K (1994) "Reflective practice", *Nursing Standard*, 8 (39): 49–56.

Baldwin, R and Ransom WH (1978) "The Integrity of Trussed Rafter Roofs", *BRE Current Paper 83/78*. Garston: BRE.

Bate, SCC (1974) "Report on the failure of the roof beams at Sir John Cass's foundation and Red Coat Church of England Secondary School, Stepney", *BRE Current Paper 58/74*. Garston: BRE.

Barrows, HS and Pickell, GC (1991) *Developing Clinical Problem-Solving Skills.* London: WW Norton & Co.

Bech-Andersen, J (1995) *The Dry Rot Fungus and other fungi in houses* (5th edition). Vejdammen, Denmark: Hussvamp Laboratoriets Forlag.

Bell, J (1999) *Doing Your Research Project* (3rd edition). Buckingham: Open University Press.

Beukel van den, A (1991) "Linking building pathology and quality assurance input", *Management, Quality and Economics in Building* (Edited by P. Brandon). London: E&FN Spon, London.

Bonshor, RB and Eldridge, LL (1974) *Graphical Aids for Tolerances and Fits: Handbook for Manufacturers, Designers and Builders.* London: HMSO.

Bonshor, RB (1977) "Jointing specification and achievement: A BRE Survey", *Current Paper 28/77*. Garston: BRE.

Bonshor, RB (1980) *Architects' Journal*, Vol. 171, No. 18, 881–885.

Bonshor, RB and Bonshor, LL (1996) *Cracking in Buildings*, Construction Research Communications Ltd, London.

Boussabaine, A (2004) *Whole Life-cycle Costing: Risk and Risk Responses.* Oxford: Blackwell Publishing.

Bowles, G and Kelly, JR (2005) "Value and risk management", *D19CV9 Course Notes for MSc Construction Management programme*, Heriot-Watt University, Edinburgh.

Brand, S (1994) *How Buildings Learn*. New York: Viking.

BRE (1971) *Digest 35 – Shrinkage of natural aggregates in concrete*.

BRE (1975) *Digest 177 – Decay and conservation of stone masonry*.

BRE (1979a) *Digest 228 – Estimation of moisture and thermal movements and stresses (Part 2)*.

BRE (1979b) *Digest 223 – Wall cladding: designing to minimise defects due to inaccuracies and movement*.

BRE (1979) *Digest 235 – Fixings for non-loadbearing precast concrete cladding panels*.

BRE (1982) *BRE Annual Report 1981/82*, The Station, Garston.

BRE (1983a) *Digest 258 – Alkali aggregate reactions in concrete*.

BRE (1983b) *Digest 268 – Common defects in low-rise buildings*.

BRE (1983c) *Digest 269 – The selection of natural building stone*.

BRE (1983c) *Digest 251 – Assessment of damage to low-rise structures*.

BRE (1983d) *Digest 277 – Built in cavity wall insulation for housing*.

BRE (1983e) *Digest 270 – Condensation in insulated domestic roofs*.

BRE (1984) *Digest 290 – Loads on roofs from snow drifting against vertical obstructions and valleys*.

BRE (1986) *Digest 312 – Flat roof design: the technical options*.

BRE (1990) *IP 15/90 – Defects in local authority housing: results of a building problems* survey.

BRE (1992) *Digest 370* – Control of lichens, moulds and similar growths.

BRE (1993) *Digest 299* – Dry rot: its recognition and control.

BRE (1996) *Good Building Guide 21 – Joist hangers*.

BRE (1997) *Good Repair Guide 12 – Wood rot: assessing and treating decay*.

BRE (2000a) *GRG 27 Part 1 – Cleaning external walls of buildings: cleaning methods*.

BRE (2000b) *GRG 27 Part 2 – Cleaning external walls of building: removing dirt and stains*.

BRE (2000c) *Digest 329* – Installing wall ties in existing construction.

BRE (2002a) "Service life assessment: benefits and mechanisms", Web presentation by D. Richardson.

BRE (2004) *Digest 330*: Alkali silica reaction in concrete (Parts 1 and 2).

BRE (2005) *Special Digest 1: Concrete in aggressive ground* (3rd edition).

Brookes, A and Meijis, M (2006) *Cladding of Buildings* (4th edition). London: Taylor & Francis.

BSI (1969) *CP 116: The structural use of precast concrete*, British Standards Institution (BSI), London.

BSI (1970) *BS 743: Specification for materials for damp proof courses*.

BSI (1971) *BS 2871-1: Copper tubes for water, gas and sanitation*, BSI, London.

BSI (1972) *CP 101: Foundations and substructures for non-industrial buildings not exceeding four storeys*, BSI, London.

BSI (1976) *BS 5390: Code of practice for stone masonry*.

BSI (1977) *BS 3837: Specification for expanded polystyrene boards*.

BSI (1978a) *BS 187: Calcium silicate (flintlime) bricks*.

BSI (1978b) *BS 6073: Precast Concrete Masonry units – Specification for precast concrete masonry units.*

BSI (1978c) *BS 5385-2: Code of practice for wall tiling: external ceramic wall tiling and mosaics.*

BSI (1978d) *BS 1243: Specification for metal ties for cavity wall construction.*

BSI (1978e) *BS 5606: Code of practice for accuracy in building.*

BSI (1979) *BS 5669: Specification for wood chipboard and methods of test for particle board.*

BSI (1980) *BS 4027: Sulphate-resisting Portland cement.*

BSI (1981) *BS 6206: Specification for impact performance requirements for flat safety glass and safety plastics for use in buildings.*

BSI (1982) *BS 6262: Code of practice for glazing of buildings.*

BSI (1983) *BS 5642: Copings of precast concrete.*

BSI (1984a) *BS 5268: Code of practice for the structural use of timber.*

BSI (1984b) "Methods for assessing exposure to wind-driven rain", *BS DD 93.* London: British Standards Institution.

BSI (1985a) *BS 5628-3: Code of practice for the structural use of masonry: materials and components, design and workmanship.*

BSI (1985b) *BS 5618: Thermal insulation of cavity walls (with masonry or concrete inner and outer leaves) by filling with urea-formaldehyde (UF) foam systems.*

BSI (1985c) *BS 8208: Guide to assessment of suitability of external cavity walls for filling with thermal insulants. Part 1: Existing traditional cavity construction.*

BSI (1985d) *BS 3921: Specification for clay bricks.*

BSI (1985e) *BS 6577: Specification for mastic asphalt for building (natural rock asphalt aggregates) [withdrawn].*

BSI (1986) *BS 8004: Code of practice for foundations.*

BSI (1987a) *BS 8110: The structural use of concrete, code of practice for design and construction.*

BSI (1987b) *BS 8201: Code of practice for flooring of timber, timber products and wood based panel products.*

BSI (1990a) *BS 8102: Code of practice for protection of structures against water from the ground.*

BSI (1990b) *BS 402: Specification for plain tiles and fittings.*

BSI (1990c) *BS 473: Specification for concrete roofing tiles and fittings.*

BSI (1991a) *BS 5262: Code of practice for external rendered finishes.*

BSI (1991b) *BS 5958–1: Code of practice for control of undesirable static electricity. General considerations.*

BSI (1994) *BS 747: Roofing felts.*

BSI (1998) *BS 8218: Code of practice for mastic asphalt roofing.*

BSI (1999) *BS 5930: Code of practice for site investigations.*

BSI (2000a) *ISO 15686-1:2000: Buildings and constructed assets – Service life planning,* Part 1: General Principles. London: British Standards Institution.

BSI (2000b) *ISO 15686-2:2000 Buildings and constructed assets – Service life planning,* Part 2: Service life prediction procedures. London: British Standards Institution.

BSI (2000c) *BS 8297: Code of practice for design and installation of non-loadbearing precast concrete panels.*

BSI (2001–2004) BS 8204-1 to 7 Screeds, bases and insitu floorings.

BSI (2002a) "Risk management vocabulary – Guidelines for use in standards", *Guide 73*, British Standards Institution, London.

BSI (2002b) *BS 5250: Control of condensation in buildings*.

BSI (2002c) *BS EN 13788: Hygrothermal performance of building components and building elements – Internal surface temperature to avoid critical surface humidity and interstitial condensation – Calculation methods*.

BSI (2003a) *BS 7543: Guide to durability of buildings, and building elements, products* and components. London: British Standards Institution.

BSI (2003b) *BS EN 14411: Ceramic tiles. Definitions, classifications, characteristics and marking* (Supersedes BS 6431-1:1983).

BSI (2003c) *BS 6229: Flats roofs with continuously supported coverings*.

Burkinshaw, R and Parrett, M (2003) *Diagnosing Damp*. London: RICS Books.

Carillion Services (2001) *Defects in Buildings: Symptoms, Investigation, Diagnosis and Care* (3rd edition). London: The Stationary Office.

Carper, KL (Editor)(1989) *Forensic Engineering*. New York: Elsevier Publishing Co.

CIB (1993) "Building pathology: A state-of the-art report", *CIB Report Publication 155*, CIB Working Commission W86, June 1993. Holland: International Council for Building Research and Information.

CIB (2003a) "Introduction to CIB". Available online at: http://www.cibworld.nl/pages/gi/Default.html.

CIB (2003b) "2nd International symposium on building pathology, durability and rehabilitation: *Learning from Errors and Defects in Building*". Available online at: http://www-ext.lnec.pt/cib_symposium_lisboa03/eng/main.htm.

CIB (2003c) "Working commission W086: Building pathology". Available online at: http://www.cibworld.nl/pages/gi/Default.html.

CIBSE (1986) *CIBSE Guide: Volume A*. London: The Chartered Institution of Building Services Engineers.

Committee on Damp Indoor Spaces and Health (2004) *Damp Inoor Spaces and Health*, Board on Health Promotion and Disease Prevention, Institute of Medicine of the National Academies. Washington DC: The National Academies Press.

Communities Scotland (2004) "Housing and health in Scotland", *Scottish House Condition Survey 2002*, SHCS Working Paper No. 4. Edinburgh: Scottish Executive.

Construction (1981) *No. 36*, pp. 29–30. London: BM Publications.

Construction (1983) *No. 42*, p. 49. London: BM Publications.

Cook, GK and Hinks, AJ (1992) *Appraising Building Defects*. London: Longman.

Cottrell, S (2005) *Critical Thinking Skills: Developing Effective Analysis and Argument*. Basingstoke: Palgrave Macmillan.

Cowan, J (1998) *On Becoming an Innovative University Teacher*. Buckingham: SRHE & Open University Press.

Croker, A (1990) *Building Failures – Recovering the Cost*. Oxford: BSP Professional Books.

Davison, J and Davison, PS (1967) *An Introduction to Spectrometry*. Dundee, UK: William Kidd & Sons Ltd.

Day, G (2003) "Opinion", *The Times Higher Education Supplement*, October 31, 2003, p. 15.

deBono, E (2003) *Six Thinking Hats* (2nd edition). London: Penguin Books Ltd.

DES (1973) *Report on the Collapse of the Roof of the Assembly Hall of the Camden School for Girls*. London: HMSO.

Diamant, RME (1977) *Insulation Deskbook*, Heating & Ventilating Publications Ltd, Croydon.

Dickinson, P and Thornton, N (2004) *Cracking and Building Movement*. London: RICS Books.

Department of the Environment (1973) *Construction No. 6*, p. 44. Croydon: Property Services Agency.

Department of the Environment (1979) *Construction No. 31*, pp. 20–21. Croydon: Property Services Agency.

Department of the Environment (1981) *Flat Roofs Technical Guide*. Croydon: Property Services Agency.

Department of the Environment (1986) *Asbestos Materials in Buildings* (2nd edition). London: HMSO.

Douglas, J (1995) Basic diagnostic chemical tests for building surveyors, *Structural Survey*, 13 (3): 22–27. MCB University Press.

Douglas, J (2006) *Building Adaptation* (2nd edition). Oxford: Butterworth-Heinemann.

Douglas, J and McEwen, IJ (1998) "Defect analysis – case study involving chemical analysis", *Construction and Building Materials*, Vol. 12, 259–267. Oxford: Elsevier Science Ltd.

Douglas, J and Singh, J (1995) "Investigating dry rot in buildings", *Building Research and Information*, Vol. 23, No. 6, 345–352. London: E&FN, Spon.

Duffy, F (1993) "Measuring building performance", *Facilities*, Vol. 8, 17–20. Bradford: MCB University Press.

Dupont™ Tyvek® (2005) *Providing protection in construction: Volume 1: Roofs*. Clevedon Dupont™ Tyvek®.

Egan, J (1998) *Rethinking Construction*. London: Department of the Environment, Transport and the Regions.

Eldridge, HJ (1974) *Common Defects in Buildings*. London: HMSO.

Endean, KF (1995) *Investigating Rainwater Penetration of Modern Buildings*. Aldershot: Gower.

Fielding, J (1987) "A kind of detective story", *Building Research & Practice (Special Feature: Building Pathology – the study of defects in buildings)*, Vol. 14, No. 2, March/April 1987. London: E&FN Spon.

Freeman, I (1987) "Deciding objectives", *Building Research & Practice (Special Feature: Building Pathology – the study of defects in buildings)*, Vol. 14, No. 2, March/April 1987. London: E&FN Spon.

Friedman, D (2006) *Advanced Home Inspection Methodology – Developing your X-Ray Vision (A Promotion Theory for Forensic Observation of Residential Construction)* (6th edition). Available online at http://www.inspect.ny.com/structure/x-ray.htm.

Garrand, C (2001) *HAPM Guide to Defect Avoidance*. London: Spon Press.

Garratt, J and Nowak, F (1991) *Tackling Condensation: A guide to the causes of, and remedies for, surface condensation and mould in traditional housing*, BRE BR Report 174. Garston: Building Research Establishment.

GB Geotechnics Ltd (2001) "Non-destructive investigation of standing structures", *Technical Advice Note 23*. Edinburgh: Historic Scotland.

Gigerenzer, G, Todd, PM and ABC Research Group (1999) *Simple Heuristics That Make Us Smart*. Oxford: Oxford University Press.

Gladwell, M (2005) *Blink (The Power of Thinking without Thinking)*, Penguin Books Ltd, London.

Graves, HM and Phillipson, MC (2000) *Potential implications of climate change in the built Environment*. London: BRE & Construction Communications Ltd.

HAPM Ltd (1991) *Defects Avoidance Manual: New Build*. Garston: Building Research Establishment.

HAPM Ltd (2000) *Life Component Manual* (2nd edition). London: E&FN Spon.

HAPM Ltd (2001) *HAPM Guide to Defect Avoidance*. London: Spon Press.

Harris, SY (2001) *Building Pathology: Deterioration, Diagnostics, and Intervention*. New York: John Wiley & Sons Inc.

Harrison, HW (1996) *Roofs and Roofing: Performance, diagnosis, maintenance, repair and the avoidance of defects*, BRE Building Elements series. London: Construction Research Communications Ltd.

Harrison, HW and de Vekey, RC (1998) *Walls, windows and doors: Performance, diagnosis, maintenance, repair and the avoidance of defects*, BRE Building Elements series. London: Construction Research Communications Ltd.

Harrison, HW and Trotman, PM (2000) *Building services: Performance, diagnosis, maintenance, repair and the avoidance of defects*, BRE Building Elements series. London: Construction Research Communications Ltd.

Harrison, HW and Trotman, PM (2002) *Foundations, basements and external works: Performance, diagnosis, maintenance, repair and the avoidance of defects*, BRE Building Elements series. London: Construction Research Communications Ltd.

Hart, JM (1991) "A practical guide to the use of thermography for building surveys", BRE *Report BR 176*, Building Research Establishment.

Haverstock, H (1999) *Building Design Easibrief* (3rd edition). Kent: Miller Freeman.

Heckroodt, RO (2002) *Guide to the Deterioration and Failure of Building Materials*, Thomas Telford Publishing, London.

Hinks, AJ and Cook, GK (1997) *The Technology of Building Defects*. London: E & FN Spon.

Hollis, M (2005) *Surveying Buildings* (5th edition). London: RICS Books.

Houghton-Evans, RW (2005) *Built Well? A forensic approach to the prevention, diagnosis and cure of building defects*. London: RIBA Enterprises Ltd.

House Condition Surveys Team (2004) *Scottish House Condition Survey 2002* (Main Report). Edinburgh: Communities Scotland.

House, JM and Kelly, GE (2000) "An overview of building diagnostics", *National Conference on Building Commissioning May 3–5 2000*, National Institute of Standards and Technology, Washington DC.

Housecroft, CE and Constable, EC (1997) *Chemistry: An Integrated Approach*. Essex., UK: Addison Wesley Longman Ltd.

Howieson, S (2005) *Housing and Asthma*. London: Taylor & Francis.

Howell, J (1994) "Diagnosis of rising damp and specification of remedial damp-proofing treatments", *RICS Focus for Building Surveying Research*, The Royal Institution of Chartered Surveyors, London.

Howell, J (1995) "Moisture measurement in masonry: Guidance for surveyors", *COBRA 95*, RICS Construction and Building Research Conference, at

Heriot-Watt University, 8–9 September 1995. London: The Royal Institution of Chartered Surveyors.

Howell, J (1996) "Rising damp – the dry facts and fiction", *The Daily Telegraph* (Property section), London, p. 37.

HSE (2003a) *Five steps to risk assessment* (Leaflet), Health & Safety Executive. Available online from http://www.hse.gov.uk/pubns/indg163.pdf

Johnson, H (1999) *International Book of Trees*. London: Mitchell Beazley.

Kahney, H (1993) *Problem Solving: Current Issues* (2nd edition). Buckingham: Open University Press.

Kahneman D, Slovic, P and Tversky, A (Editors) (1982) *Judgment Under Uncertainty: Heuristics and Biases*. Cambridge University Press, Cambridge, England.

Kaminetzky, D (1992) *Design and Construction Failures: Lessons from Forensic Investigations*. New York: McGraw-Hill Publishing Co.

Latham, M (1994) *Constructing the Team*, Final Report. London: HMSO.

Lacy, RE (1976) *Driving Rain Index*. London: HMSO.

Lutz, KF, Jones, KD, Kendall, J (1997) "Expanding the praxis debate: contributions to clinical enquiry", *Nursing Science*, Vol. 20, No. 2, 23–31.

Manketlow, J (2003) *Mind Tools* (E-book). London: Mind Tools Ltd.

Marshall, D, Worthing, D and Heath, R (2003) *Understanding Housing Defects* (2nd edition) London: Estates Gazette.

Massari, G and Massari, I (1993) *Damp Buildings, Old and New* (Translated by C. Rockwell). Rome: ICCROM.

McAleese, R (2003) "Personal communication with J. Douglas", November 2003, Heriot-Watt University, Edinburgh.

Moon, J (1999) *Reflection in Learning and Professional Development*. London: Routledge Falmer.

Moore, JFA (1984) "The use of glass-reinforced cement in cladding panels", BR 49. Garston, Watford, Herts: Building Research Establishment.

NBA (1983) *Common Building Defects: Diagnosis and Remedy*. London: Construction Press.

NBA (1985) *Maintenance Cycles & Life Expectancies of Building Components & Materials: A guide to data and sources*. London: National Building Agency.

Neville, AM and Brooks, JJ (1987) *Concrete Technology*. Harlow: Longman.

NHBC (2005) *Technical Standards Vols 1 and 2*. London: National House-Building Council.

Nixon, PJ (1978) *Chemistry and Industry*, 5, 160–164.

NMAB (1982) "Conservation of historic stone buildings and monuments", *Report of the Committee on Conservation of Historic Stone Buildings and Monuments*. Washington DC: National Academy Press.

Noy, E and Douglas, J (2005) *Building Surveys and Reports* (3rd edition). Oxford: Blackwell Publishing, Oxford.

O'Callahan, N (2005) "The use of expert practice to explore reflection", *Nursing Standard*, Vol. 19, No. 39.

Odom, JD and Dubose, G (1999) *Commissioning Buildings in Hot, Humid Climates: Design and Construction Guidelines*. Orlando, FL: CH2M HILL and Fairmont Press Ltd.

Odom, JD and Dubose, G (2002) *Preventing Moisture and Mold Problems in Hot, Humid Climates: Design and Construction*. Orlando, FL: CH2M HILL and Fairmont Press Ltd.

ODPM (2003) *Empty Homes: Temporary Management, Lasting Solutions*, "A Consultation Paper", October 2003, Office of the Deputy Prime Minster. London: HMSO.

OGC (2003) *Procurement Guide 07: Whole-life costing and cost management (Achieving Excellence in Construction)*, Office of Government Commerce. Available online from www.ogc.gov.uk.

Palfreyman, JW and Urquhart, D (2002) "The environmental control of dry rot", *Technical Advice Note 24*, Technical Research & Conservation Division. Edinburgh: Historic Scotland.

Parnham, P (1996) *Prevention of Premature Staining of New Buildings*. London: Chapman & Hall.

Parnham, P and Rispin, C (2001) *Residential Property Appraisal*. London: E&FN Spon.

Plous, S (1993) *The Psychology of Judgement and Decision Making*. New York: McGraw-Hill.

Potter, PA and Perry, GA (2001) *Fundamentals of Nursing* (5th edition). St Louis: Mosby.

Princes Risborough Research Laboratory (1982) "The movement of timber", *Technical Note 28*. Garston: Princes Risborough Laboratory.

Puller-Strecker, P (1990) "Building defects: Are they avoidable?", Keynote Speech, IN *Building Maintenance & Modernisation Worldwide*, Vol. 1 (Edited by Quah, Lee Kiang), Proceedings of the International Symposium on Property Maintenance and Modernisation, 7–9 March 1990, Singapore. London: Longman.

Pye, P and Harrison, HW (1997) *Floors and Flooring: Performance, diagnosis, maintenance, repair and the avoidance of defects*, BRE Building Elements series. London: Construction Research Communications Ltd.

Race, P (2004) *Evidencing Reflection – putting the "w" into reflection* (ESCALATE Learning Exchange). Available online at: http://escalate.ac.uk/exchange/Reflection [accessed 26 February 2004].

Raftery, J (1994) *Risk Analysis in Project Management*. E&FN Spon, London.

Research, Analysis & Evaluation Division (2003) *English House Condition Survey 2001* (Main Report). London: Office of the Deputy Prime Minister.

Richardson, BA (2001) *Defects and Deterioration in Buildings* (2nd edition). London: Spon Press.

Ridout, B (1999) *Timber Decay in Buildings: The conservation approach to treatment*, English Heritage and Historic Scotland. London: E&FN Spon.

Rose, WB (2005) *Water in Buildings: An Architect's Guide to Moisture and Mold*. New York: John Wiley & Sons.

Rougier, P and Lefly, M (1995) "The diagnosis of building defects by computer", *Construction Papers, No. 34* (Edited by P. Harlow). Englemere, England: Chartered Institute of Building.

Rushton, T (1992), "Understanding why buildings fail and using 'Heir' methodology in practice", *Latest Thinking on Building Defects in Commercial Buildings: Their Diagnosis, Causes and Consequences*, One-day Conference, Wednesday 15 July 1992, Henry Stewart Conference Studies, London.

Sackett, DL, Rosenberg, WMC, Gray, JAM, Haynes, RB and Richardson, WS (1996) "Evidenced-based medicine: What it is and what it isn't", *British Medical Journal*, January 1996, 312, 71–72.

Savory, JB and Carey, JK (1975) *Timber Trades Journal, Wood Treatments Supplement 295*, (5171), pp. 12–13.

Schon, DA (1995) *The Reflective Practitioner: How Professionals Think in Action* (2nd edition). New York Basic: Books Inc. Publishers.

Scottish Office and Scottish Homes (1989) *Housing Maintenance Kit*. Edinburgh: HMSO.

SearchSecurity.com (2005) *Definitions – heuristics*. Available online from http://searchsecurity.techtarget.com/sDefinition/0,sid14_gci876400,00.html [accessed 17/02/06].

Smith, JL (1987) "Frameworks for defects diagnosis", *Building Research & Practice (Special Feature: Building Pathology – the study of defects in buildings)*, Vol. 14, No. 2, March/April 1987. London: E&FN Spon.

Son, T and Yuen, G (1993) *Building Maintenance Technology*. London: Macmillan.

Thaumasite Expert Group (1999) "The thaumasite form of sulphate attack: Risks, diagnosis, remedial work and guidance on new construction", *Report of the Thaumasite Expert Group*, January 1999. London: Department of the Environment, Transport and the Regions.

Tomlinson, MJ, Driscoll, R and Burland, JB (1978) *The Structural Engineer*, Vol. 56A, No. 6, pp. 161–173.

Trotman, P and Sanders, C and Harrison, H (2004) *Understanding Dampness – Effects, Causes, Diagnosis and Remedies*, BR466. Garston: Building Research Establishment.

Vegoda, V (1993) *A Practical Guide to Residential Surveys*, Need to Know Series No. 2. London: Lark Productions Ltd.

Ward, WH (1948) "The effect of vegetation on the settlement of structures", *Proceedings Conference on Biology and Civil Engineering*. London: ICE.

Warszawski, A (1999) *Industrialized and Automated Building Systems* (2nd edition). London: E&FN Spon.

Watt, D (1999) *Building Pathology: Principles and Practice*. Oxford: Blackwell Science.

Weatherproofing Advisors Ltd (2004) "Defects associated with organically coated steel sheeting", *Technical Guide*.

Wheeler, I (2002) "Glass failure: The unacceptable risk", *Environmental Health Journal*, July 2002.

Wilkinson, J (1999) "Implementing reflective practice", *Nursing Standard*, 13 (21), 36–40.

Wood, B (2003) *Building Care*. Oxford: Blackwell Publishing.

Further reading

Atkinson, MF (2000) *Structural Defects Reference Manual for Low-Rise Buildings*. Oxford: Taylor and Francis.

Barrett, K (2005) *Defective Construction Work*. Oxford: Blackwell Publishing.

Bliss, S (Editor)(1997) *Troubleshooting Guide to Residential Construction*. Williston, VT: Journal of Light Construction.

Brantley, LR and Brantley, RT (1996) *Building Materials Technology (Structural Performance and Environmental Impact)*. New York: McGraw-Hill Inc.

Bravery, AF, Berry, RW, Carey, JK and Cooper, DE (2005) *Recognising Wood rot and Insect Damage in Buildings* (3rd edition). Garston: Building Research Establishment.

BRE (2002b) *Special Digest 2: HAC in the UK: Assessment, durability management, maintenance, repair and refurbishment.*

British Steel (1996) *The Prevention of Corrosion on Structural Steelwork.* Cleveland: British Steel Publications.

Brown, SA (2000) *Communication in the Design Process.* London: Routledge Falmer.

Brungs, MP and Sugeng, XY (1995) "Some solutions to the nickel sulphide problem in toughened glass", *Glass Technology*, Vol. 36, No. 4, August 1995.

Bryan, AJ (1989) "Movements in buildings", CIOB Technical Information Service. Englemere, England: Chartered Institute of Building.

CABE (2006) *The Cost of Bad Design.* London: Commission for Architecture and the Built Environment.

Cash, CG (2003) *Roofing Failures.* Oxford: Taylor and Francis.

Clarke, LA and Thaumasite Expert Group (2002) *The Thaumasite Expert Group Report: Review after three years experience.* London: Office of the Deputy Prime Minister.

Constable, A and Lamont, C (2006) *Case in Point – Building Defects.* London: RICS Books.

Construction Audits Ltd and HAPM (1993) *Defects Avoidance Manual: New Build.* Garston: Building Research Establishment.

Cox, RH, Kempster, JA and Bassi, R (1993) "Survey of performance of organic-coated roof sheeting", *BRE Report BR 359.* Garston: Building Research Establishment.

Curtin, WG (1993) "Causes of defects", *Appraisal and repair of building structures: Introductory guide* (Edited by R. Holland *et al.*). London: Thomas Telford.

Curtin, W and Parkinson, E (1989) *Structural Appraisal and Restoration of Victorian Buildings,* Conservation and Engineering Structures. London: Thomas Telford.

Dainty, A, Moore, D and Murray, M (2005) *Communication in Construction: A Concise Guide for the Construction Industry.* London: RoutledgeFalmer.

Doran, D and Cockerton, C (Editors) (2007) *Principles and Practice of Testing in Construction.* Dunbeath, Caithness, Scotland: Whittles Publishing.

Duell, J and Lawson, F (1977) *Damp Proof Course Detailing.* London: The Architectural Press.

Erdly, JL and Schwartz, TA (Editors) (2000) "Building Façade Maintenance, Repair and Inspection", *ASMT STP 1444.* West Conshohocken, PA: ASMT.

Glover, P (2006) *Building Surveys* (6th edition). Oxford: Butterworth-Heinemann.

Groner, RF (1983) "The role of Heuristics in Models of Decision", in *Decision Making under Uncertainty* (Edited by RW Scholz), Advances in Psychology 16, North-Holland, Oxford.

Elder, B (2005) *Building Pathology – Residential Dwellings.* Reading: College of Estate Management.

Feld, J and Carper, KL (1996) *Construction Failure* (2nd edition). Canada: John Wiley & Sons.

Handisyde, CC (1984) *Everyday Details.* London: Architectural Press.

HAPM Ltd (2000) *Life Component Manual* (2nd edition). London: E&FN Spon.

Holland, R, Montgomery-Smith, B and Moore, J (Editors)(1990) *Appraisal and Repair of Building Structures*. London: Thomas Telford Services Ltd.

Hollis, M (1990) *Cavity Wall Tie Failure*. London: Estates Gazette.

Holmes, S and Wingate, M (1997) *Building with Lime: A practical introduction*. London: Intermediate Technology Publications.

House Condition Surveys Team (2005) *Housing and Disrepair in Scotland: Analysis of the 2002 Scottish House Condition Survey*. Edinburgh: Communities Scotland.

Hoxley, M (2002) *Construction Companion to Building Surveys*. London: RIBA Publications.

Hughes, P (1986) "The need for old buildings to 'Breathe' ", *Information Sheet 4*. London: SPAB.

HSE (1990) "Evaluation and inspection of buildings and structures", *Health and Safety series booklet HS(G)58*. London: HMSO.

HSE (2003b) *Managing health and safety: Five steps to success* (Booklet), Health & Safety Executive. Available online from http://www.hse.gov.uk/pubns/indg275.pdf.

Institution of Structural Engineers (1992) *Appraisal of Building Structures*. London: Institution of Structural Engineers.

Institution of Structural Engineers (1995) *Expert Evidence – A Guide for Expert Witnesses and Their Clients*. London: SETO Ltd.

Institution of Structural Engineers (2000) *Subsidence of low-rise buildings* (2nd edition). London: Institution of Structural Engineers.

Kudder, RJ and Erdly, JL (Editors) (1998) "Water Leakage Through Building Facades", *ASMT STP 1314*. West Conshohocken, PA: ASMT.

Lead Sheet Association (1990) *Lead Work Technical Notes 3*, "Condensation – a problem . . . even for lead". London: Lead Sheet Association.

Macmillan, S (Editor) (2004) *Designing Better Buildings*. London: Spon Press.

Matulionis, RC and Freitag, JC (Editors) (1990) *Preventive Maintenance of Buildings*. New York: Van Nostrand Reinold.

Mayer, P and Wornell, P (Editors) (1999) *HAPM Workmanship Checklists*, HAPM Publications Ltd. London: E&FN Spon.

MCRMA (1991) "Curved sheeting manual", *Technical Paper No. 2*, Metal Cladding & Roofing Manufacturers Association (MCRMA).

MCRMA (1992) "Secret fix roofing guide", *Technical Paper No. 3*, MCRMA.

MCRMA (1992) "Fire and external steel-clad walls – guidance", Notes to the Revised Building Regulations 1992, *Technical Paper No. 4*, MCRMA.

MCRMA (1992) "Metal wall cladding: Detailing guide", *Technical Paper No. 5*, MCRMA.

MCRMA (1994) "Fire design of steel sheet clad external walls for building – Construction performance standards and design", *Technical Paper No. 7*, MCRMA.

MCRMA (1994) "Acoustic design guide for metal roof and wall cladding systems", *Technical Paper No. 8*, MCRMA.

MCRMA (1995) "Composite roof and wall cladding panel design guide", *Technical Paper No. 9*, MCRMA.

MCRMA (1995) "Profiled metal cladding for roofs and walls – Guidance notes on the revised building regulations 1995, Parts L & F," *Technical Paper No. 10*, MCRMA.

MCRMA (1996) "Profiled metal roofing design guide", *Technical Paper No. 6 (Revised Edition)*, MCRMA.

MCRMA (1997a) "Recommended good practice for daylighting in metal clad buildings", *Technical Paper No. 1 (Revised Edition)*, MCRMA.

MCRMA (1997b) "Flashings for metal roof and wall cladding: Design, detailing and installation Guide", *Technical Paper No. 11*, MCRMA.

MCRMA (2000) "Fasteners for metal roof and wall cladding: Design, detailing and Installation Guide", *Technical Paper No. 12*, MCRMA.

MCRMA (2002) "Guidance for the design of metal and cladding roofing to comply with approved document", *Technical Paper No. 14*, MCRMA.

MCRMA (2004) "Guidance for the effective sealing of end lap details in metal roof constructions", *Technical Paper No. 16*, MCRMA.

McDonald, S (Editor) (2002) *Building Pathology: Concrete*. Oxford: Blackwell Publishing.

McKenna, H, Cutliffe, J and McKenna, P (1999). "Evidence based practice – demolishing some myths", *Nursing Standard*, 14 (16): 39–42.

McKibbon, KA (1998) "Evidence based practice", *Bulletin of the Medical Library Association*, 86 (3): 396–401.

Mika, SJ and Desch, S (1992) *Structural Surveying* (2nd edition). London: Macmillan.

Mills, E (Editor) (1994) *Building Maintenance and Preservation* (2nd edition). London: Architectural Press.

Murdoch, R (1990) *The Lead Sheet Manual – A Guide to Good Building Practice: Volume 1 Lead Sheet Flashings*. London: Lead Sheet Association.

Murdoch, R (1992) *The Lead Sheet Manual – A Guide to Good Building Practice: Volume 2 Lead Sheet Roofing and Cladding*. London: Lead Sheet Association.

Murdoch, R (1993) *The Lead Sheet Manual – A Guide to Good Building Practice: Volume 3 Lead Sheet Weatherings*. London: Lead Sheet Association.

Oliver, A (1997) *Dampness in Buildings* (2nd edition revised by J. Douglas & J.S. Stirling). Oxford: Blackwell Science.

Oxley, TA and Gobert, EG (1994) *Dampness in Buildings (Diagnosis, Instruments, Treatment)* (2nd edition). Oxford: Butterworth-Heinemann.

Palfreyman, JW, Smith, D and Low, G (2002) "Studies of the domestic dry rot fungus *Serpula Lacrymans* with Relevance to the management of decay in buildings", *Research Report*, Technical Research & Conservation Division. Edinburgh: Historic Scotland.

Preiser, WF and Vischer, J (Editors) (2004) *Assessing Building Performance*. New York: Butterworth-Heinemann.

Richardson, C (1987) *AJ Guide to Structural Surveys*. London: Architectural Press.

Richardson, C (1996) "When the earth moves", *Architects Journal*, 22 February 1996. London.

Richardson, C (2000) "Moving structures", *Architects Journal*, 14 September 2000. London.

RICS (1995) *Flat Roof Covering Problems – RICS Guidance Note*. London: RICS Books.

RICS (1997) *The Mundic Problem – RICS Guidance Note* (2nd edition). London: RICS Books.

RICS (2002) *The Mundic Problem Supplement – Stage 3 Expansion Testing*. London: RICS Books.

Robson, P (1991) *Structural Appraisal of Traditional Buildings*. Shaftsbury: Donhead Publishing.

Ross, P (2001) *Appraisal and Repair of Timber Structures*. London: Thomas Telford Services Ltd.

Rushton, T (2006) *Understanding Hazardous and Deleterious Materials*. London: RICS Books.

Scott, G (1976) *Building Disasters and Failures*. Lancaster: Construction Press.

Smith, L (1985) *Investigating Old Buildings*. London: Batsford Academic and Educational.

Watts and Partners (2007) *Watts Pocket Handbook 2007*. London: Watts & Partners.

Some useful internet sites

The number of Internet sites relating to construction and surveying is growing rapidly. The following list, therefore, is a sample of some of the most prominent sites available that address building failures and related issues.

www.ashrae.org/ (website of the American Society of Heating, Refrigeration and Air-Conditioning.

Engineers – for information on commissioning building systems, etc.).

www.bre.co.uk (website of the Building Research Establishment).

www.buildingpreservation.com/ (Graham Coleman's informative website on dampness and timber treatment).

www.cbppp.org.uk/cbpp (for information on the Construction Best Practice Programme).

www.cibworld.nl (website of the International Council for Research and Innovation in Building and Construction – CIB for short – based in Holland).

www.cisti.nrc.ca/irc (website of the Canadian Institute for Research in Construction, which contains and extensive list of useful articles on a whole range of building problems).

www.clasp.gov.uk (for full details of the various types of CLASP system buildings – including case studies of refurbished properties built using this method).

www.concreterepair.org.uk (website of the Concrete Repair Association).

www.dti.gov.uk (for business-related information on the construction industry).

www.dryrotdogs.co.uk (Hutton + Rostron website for technical information on dry rot and other moisture-related problems).

www.fogit.co.uk (Rentokil's property care website).

www.housedustmite.org.uk (for useful information on the risks and remedies associated with house dustmite infestation in dwellings).

www.mcrma.co.uk (website of the Metal Cladding and Roofing Manufacturers Association).

www.odpm.gov.uk (website of the Office of the Deputy Prime Minister, which contains information on the Building Regulations and other technical matters related to construction).

www.peci.org/library.htm (website of Portland Energy Conservation Inc containing several articles/reports on commissioning buildings and related issues).

www.safeguardchem.com/Downloads (for downloads on damp proofing and timber treatments).

www.ukonline.gov.uk (for information on a wide range of sources relating to general and local government).

www.weatherproofing.co.uk (website of Weatherproofing Advisers Ltd).

www2.cr.nps.gov/tps/briefs/presbhom.htm (website of the US National Park Service heritage preservation services containing an excellent range of articles and briefs on the inspection, appraisal and conservation of historic buildings).

Index

305·89

GARVEY: HIS WORK AND IMPACT